SpringerBriefs in Applied Sciences and Technology

Mathematical Methods

Series editor

Anna Marciniak-Czochra, Heidelberg, Germany

More information about this series at http://www.springer.com/series/11219

Kexiang Xu · Kinkar Ch. Das
Nenad Trinajstić

The Harary Index of a Graph

Kexiang Xu
Department of Mathematics,
  College of Science
Nanjing University of Aeronautics
  and Astronautics
Nanjing, Jiangsu
China

Kinkar Ch. Das
Department of Mathematics,
  College of Science
Sungkyunkwan University
Gyeonggi-do
Korea, Republic of Korea

Nenad Trinajstić
Rugjer Bošković Institute
Zagreb
Croatia

ISSN 2191-530X   ISSN 2191-5318   (electronic)
SpringerBriefs in Applied Sciences and Technology
ISBN 978-3-662-45842-6   ISBN 978-3-662-45843-3   (eBook)
DOI 10.1007/978-3-662-45843-3

Library of Congress Control Number: 2014957864

Mathematics Subject Classification: 05C12, 05C90, 05C35, 05C05, 05C07, 92E10

Springer Heidelberg New York Dordrecht London

Printed on acid-free paper

Springer-Verlag GmbH Berlin Heidelberg is part of Springer Science+Business Media
(www.springer.com)

This book is dedicated to Frank Harary (New York, NY, 1921-Las Cruces, NM, 2005), the grandmaster of graph theory and its applications.

Additionally, KX dedicate this book to his daughter Anran Xu, KCD to his daughter Kankana Das, NT to his wife Judita.

Preface

The Harary index of a graph has been independently introduced by two research groups, one in Zagreb (Croatia) and other in Bucharest (Romania), at the Symposium held in honour of Professor Frank Harary at the University of Saskatchewan, Saskatoon, Canada from September 12 to 14, 1991. This symposium was organized to celebrate the 70th birthday of Harary. The term *Harary index* was given by the Zagreb Group [1], a member of which is one of the present authors (NT), and denoted by H, while the Bucharest Group called this index as the *reciprocal distance sum index* and denoted it RDSUM [2]. However, the term Harary index nowadays is generally accepted for this molecular descriptor [3, 4].

The Harary index of a graph G is defined [1, 2] as

$$H(G) = \sum_{u,v \in V(G)} \frac{1}{d_G(u,v)}$$

where $d_G(u,v)$ denotes the distance of two vertices u, v in G and the summation goes over all unordered pairs of vertices of G.

Since its first introduction in 1991, the Harary index has attracted much attention of chemical and mathematical researchers, especially those focussing on graph theory, from all over the world. Nowadays many interesting results on Harary index have been reported in literature [3, 4]. These results range from theoretical ones such as the extremal graphs with respect to Harary index, the relation between Harary index and other topological indices of graphs and some properties of Harary

---

[1] These lines are taken from his article: The unreasonable effectiveness of mathematics in the natural sciences, Comm. Pure Appl. Math. 13 (1960) 1–14.

index, and so on, to applied ones, including its applications in pure graph theory or in mathematical chemistry. Very recently, an interesting variant of Harary index of a graph G, which is named as additively weighted Harary index, has been introduced [5]. For a graph G, the additively weighted Harary index is defined [5] as follows:

$$H_A(G) = \sum_{\{u,v\} \subseteq V(G)} \frac{d_G(u) + d_G(v)}{d_G(u,v)}$$

Moreover some mathematical results have been obtained in several recent papers (e.g., [6, 7]).

In this book we will report some properties and applications of Harary index as well as some mathematical results of additively weighted Harary index. Furthermore, in the last chapter we will propose some interesting open problems on the Harary index of graphs.

Nanjing, November 2014   Kexiang Xu
Suwon   Kinkar Ch. Das
Zagreb   Nenad Trinajstić

References

1. Plavšić D, Nikolić S, Trinajstić N, Mihalić Z (1993) On the Harary index for the characterization of chemical graphs. J Math Chem 12:235–250
2. Ivanciuc O, Balaban TS, Balaban AT (1993) Design of topological indices. Part 4. Reciprocal distance matrix, related local vertex invariants and topological indices. J Math Chem 12:309–318
3. Todeschini R, Consonni V (2000) Handbook of Molecular Descriptors. Wiley-VCH, Weinheim, pp 497–502
4. Todeschini R, Consonni V (2009) Molecular Descriptors for Chemoinformatics, Vol. I, Vol. II., Wiley-VCH, Weinheim, pp 934–938
5. Alizadeh Y, Iranmanesh A, Došlić T (2013) Additively weighted Harary index of some composite graphs. Discrete Math 313:26–34.
6. Janežič D, Miličević A, Nikolić S, Trinajstić N (2007) Graph Theoretical Matrices in Chemistry, University of Kragujevac, Kragujevac
7. Lučić B, Sović I, Plavšić D, Trinajstić N (2012) Harary matrices: definitions, properties and applications. In: Gutman I, Furtula B (eds.) Distance in molecular graphs-applications, University of Kragujevac, Kragujevac, p 3–26

Acknowledgments

The first author was supported by the NNSF of China (No. 11201227), China Postdoctoral Science Foundation (2013M530253, 2014T70512) and Natural Science Foundation of Jiangsu Province (BK20131357). The second author was supported by the Faculty research Fund, Sungkyunkwan University, 2012 and National Research Foundation funded by the Korean government with the grant No. 2013R1A1A2009341. The third author was supported by the grant from the Ministry of Science, Education and Sports of Croatia.

Contents

Figures

Tables

Chapter 1
Introduction

1.1 Short Introduction to Graph Theory

The solution of Königsberg Bridge Problem in 1736 by a great Swiss mathematician Leonhard Euler (1707–1783) gave birth to a novel subject—Graph Theory, which also made him the father of graph theory. Two hundred years later the first book on graph theory, "*Theorie der endlichen und unendlichen Graphen*" (Teubner, Leipzig), by Denes König was published in 1936. Since then graph theory has formally become a branch of mathematics. Till date, graph theory has not only been used in other branches of mathematics but has also been extensively applied in other scientific subjects such as information theory, computer science, economics, physics, chemistry, and so on.

A graph G is defined as a pair of sets $G = (V, E)$ with $E \subset V \times V$. Therefore G represents a binary relation. The graph is undirected if the binary relation is symmetric and directed otherwise. A graph G is called simple if it does not contain loops or multiple edges.

Let $G = (V(G), E(G))$ be a graph with vertex set $V(G)$ and edge set $E(G)$. For a vertex $v \in V(G)$, we denote by $N_G(v)$ the set of neighbors of v in G, and $d_G(v) = |N_G(v)|$ is called the degree of v in G. In particular, we denote by $\Delta = \Delta(G)$ the maximum degree of vertices of G, and by $\delta(G)$ the minimum degree of vertices of G. A vertex v of degree 1 is called *pendant vertex*. For a subset W of $V(G)$, let $G - W$ be the subgraph of G obtained by deleting the vertices of W and the edges incident with them. Similarly, for a subset E' of $E(G)$, we denote by $G - E'$ the subgraph of G obtained by deleting the edges of E'. If $W = \{v\}$ and $E' = \{xy\}$, the subgraphs $G - W$ and $G - E'$ are written as $G - v$ and $G - xy$ for short, respectively.

In what follows, we denote by P_n, S_n, C_n and K_n the path graph, the star graph, the cycle graph, and the complete graph with n vertices, respectively. A graph G is connected if for every two vertices $u, v \in V(G)$ there exists a (u, v)-path in G. Otherwise it is called disconnected. The maximal connected subgraphs of G are called its components. The *vertex-connectivity* is the minimum number of vertices whose

© The Author(s) 2015
K. Xu et al., *The Harary Index of a Graph*,
SpringerBriefs in Mathematical Methods, DOI 10.1007/978-3-662-45843-3_1

deletion from a connected graph disconnects it. Similarly, the *edge-connectivity* is the minimum number of edges whose deletion from a connected graph disconnects it.

It is well known that a tree is a connected graph which does not contain any cycle as a subgraph. The pendant vertex in a tree is also called a *leaf*. For two vertex-disjoint graphs G_1 and G_2, we denote by $G_1 \cup G_2$ a new graph with vertex set $V(G_1) \cup V(G_2)$ and with edge set $E(G_1) \cup E(G_2)$. In particular, we write $\overbrace{G \cup G \cdots \cup G}^{k} = kG$. Moreover, the *join* of G_1 and G_2, denoted by $G_1 \bigvee G_2$, is the graph with vertex set $V(G_1) \bigcup V(G_2)$ and edge set $E(G_1) \bigcup E(G_2) \bigcup \{u_i v_j : u_i \in V(G_1), v_j \in V(G_2)\}$. For any graph G, we denote by $\overline{G}$ the complement of G. For other notations and terminology of graph theory, the readers are referred to monographs [1, 2].

1.2 Distance in Graphs

Throughout this book we only consider undirected simple graphs (i.e., graphs without loops and multiple edges). For any two vertices $u, v \in V(G)$ of a graph G, the *distance* between them, denoted by $d_G(u, v)$, is the length of a shortest path connecting them in G (just the number of edges in this path). Furthermore, the *diameter* of graph G is the largest distance between any two vertices in G.

The distance between two vertices is a fundamental concept in pure graph theory, which is the basis of some basic definitions of graphs such as diameter, radius, eccentricity, mean distance, and so on. What is more, distance-regular graphs [3] and distance-transitive graphs [4] are two important graph classes based on the distance in graph. Some details on the distance between vertices can be found in a well-known monograph [5] and an excellent chapter [6].

Moreover, the distance between two vertices of a graph also plays an important role in applied graph theory. Particularly, it is much useful in other subjects such as mathematical chemistry. In this field many distance-based molecular descriptors have been introduced to reflect partially the structure of various molecules (e.g., [7–9]). Nowadays the distance itself and other related topics have attracted more and more attention from many mathematicians and chemists all over the world, especially those on Graph Theory or Mathematical Chemistry. For details on the distance in molecular graphs, readers can refer to four related monographs [5, 10–12].

1.3 Harary Index of a Graph

In mathematical chemistry, topological index, also known as molecular descriptor, is a single number that can be used to characterize some property of the graph of a molecule. From the 1940s to now, hundreds of topological indices of (molecular) graphs are introduced for various purposes in chemistry. Among these, topological indices of graphs that are defined on the basis of distances in graphs represent a large family of molecular descriptors [8–11].

In 1947, Wiener [13] first introduced the Wiener index when calculating the boiling point of alkanes. Probably Wiener index is the first topological index used in Mathematical Chemistry. Nowadays, the Wiener index of a graph G is defined [14] as follows:

$$W(G) = \sum_{u,v \in V(G)} d_G(u, v)$$

where the summation goes over all unordered pairs of vertices of G and hereafter this sign will have the same meaning when it is in this form. The hyper-Wiener index of a graph G, first introduced by Randić [15] in 1993, is defined as [16]:

$$WW(G) = \frac{1}{2} \sum_{u,v \in V(G)} d_G(u, v) + \frac{1}{2} \sum_{u,v \in V(G)} d_G(u, v)^2.$$

The Harary index of a graph G, denoted by $H(G)$, was introduced independently by Plavšić et al. [17], including one of the present authors (NT), and by Ivanciuc et al. [18] in 1993. Note that a version of the Harary index, the squared Harary index H^2, was first defined and tested in 1992 by Mihalić and one of the present authors [19] as:

$$H^2(G) = \sum_{u,v \in V(G)} \frac{1}{d_G(u, v)^2}.$$

The Harary index is defined [17, 18] as follows:

$$H = H(G) = \sum_{u,v \in V(G)} \frac{1}{d_G(u, v)}. \tag{1.1}$$

In this book, we always use the form in (1.1) for Harary index of a graph. The Harary index may for example be also defined in the following way. Let $\gamma(G, k)$ be the number of vertex pairs of the graph G that are at distance k. Then

$$H(G) = \sum_{k \geq 1} \frac{1}{k} \gamma(G, k). \tag{1.2}$$

For a disconnected graph G with k components $G_1, G_2, \ldots, G_k$, we have [20, 21]

$$H(G) = \sum_{i=1}^{k} H(G_i).$$

In chemical graph theory, there are several important distance-based topological indices of graphs that are extensively studied in many mathematical and chemical papers. They are Wiener polarity index [13], reciprocal complementary Wiener index [22, 23], terminal Wiener index [24, 25], Szeged index [26], and Balaban index [27, 28], etc. Wiener index and hyper-Wiener index are extensively reported in the literature [8, 9, 29–36]. Some details of mathematical properties of distance-based topological indices can be found in [37]. Mathematical properties and applications of the family of Wiener and Wiener-like indices are extensively reported in the literature [8, 9, 12, 13, 22–25, 27–44]. In this book, we only focus on the Harary and Harary-like indices of graphs.

1.4 Harary Matrix of a Graph

In the process of inventing the Harary index of a graph [17, 18], the reciprocal distance matrix [17, 18, 45] (also named as Harary matrix) played an important role.

For a connected graph G of order n, the distance matrix $D(G)$ [17, 18] is an $n \times n$ matrix (d_{uv}) where d_{uv} denotes the distance between the vertices u and v in the graph G. And the reciprocal distance matrix $RD(G)$ [17, 18] is also an $n \times n$ matrix (RD_{uv}), where RD_{uv} is $\dfrac{1}{d_{uv}}$ if $u \neq v$ for any two distinct vertices $u, v \in V(G)$, and 0 otherwise. Therefore we have

$$H(G) = \sum_{u,v \in V(G), u \neq v} RD_{uv}.$$

If we denote by $S(M)$ the sum of all entries in a square matrix M, then, from the definitions of Harary index and reciprocal distance matrix, respectively, we have

$$H(G) = \frac{1}{2}S(RD(G)). \tag{1.3}$$

For some simple (molecular) graphs the Harary index can be given in a closed form. We list below analytical formulae for some classes of simple graphs.

(i) Harary indices for chains, depicting, for example, the carbon skeletons of the n-alkanes are given by:

$$H = n \sum_{k=1}^{n} \frac{1}{k} - (n-1) \tag{1.4}$$

where n is the number of vertices (carbon atoms) in the chain (n-alkane).

(ii) Harary indices for n-cycles, depicting, for example, the carbon skeletons of n-annulenes are given by:

Fig. 1.1 Examples of the star graph on five points (S_5) and the complete graph on five points (K_5)

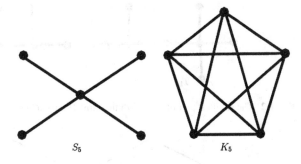

S_5 K_5

n is even,

$$H = n \sum_{k=1}^{n-2} \frac{1}{k} + 1; \tag{1.5}$$

n is odd,

$$H = n \sum_{k=1}^{n-1} \frac{1}{k}. \tag{1.6}$$

(iii) Harary indices for star graphs S_n:

$$H(S_n) = \frac{1}{4}(n+2)(n-1). \tag{1.7}$$

Example of the star graph is given in Fig. 1.1

(iv) Harary indices of complete graphs K_n

$$H(K_n) = \frac{1}{2}n(n-1). \tag{1.8}$$

The Harary index H has several interesting properties. Some of these we list below:

(i) The Harary index is not unique since there are non-isomorphic (molecular) graphs with identical Harary indices. The smallest pair of the trees with the same Harary index are given in Fig. 1.2 and the smallest pair of polycyclic graphs with the same Harary index is given in Fig. 1.3.

(ii) The Harary index appears to be a convenient measure of the compactness of the molecule. The larger the Harary index, the larger the compactness of the molecule. This is illustrated for hexanes. Their carbon skeleton are depicted as trees in Fig. 1.4. The corresponding Harary indices are as follows: **1**: 8.7000; **2**: 9.0000; **3**: 9.0833; **4**: 9.3333; **5**: 9.5000. The least compact molecule appears

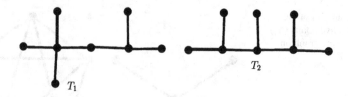

Fig. 1.2 The smallest pair of trees with the same Harary index

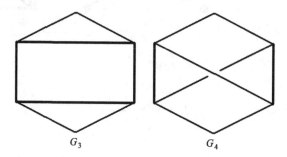

Fig. 1.3 The smallest pair of polycyclic graphs with the same Harary index

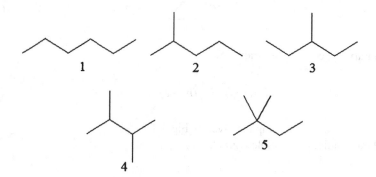

Fig. 1.4 Trees modeling carbon skeletons of isomeric hexanes

to be n-hexane ($H = 8.700$) and the most compact molecule 2,2-dimethylbutane ($H = 9.500$). The order of hexanes also follows the branching pattern of heptanes. Thus, the Harary index appears to be useful molecular descriptor for considering branching in alkane trees.

As given in [46], for a tree T as shown in Fig. 1.5, we can compute its distance matrix $D(T)$ and reciprocal distance matrix RD(T), respectively, as follows:

Fig. 1.5 A tree T

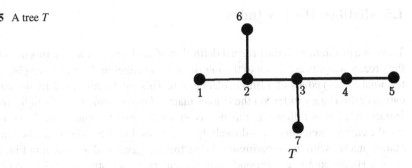

$$D(T) = \begin{pmatrix} 0 & 1 & 2 & 3 & 4 & 2 & 3 \\ 1 & 0 & 1 & 2 & 3 & 1 & 2 \\ 2 & 1 & 0 & 1 & 2 & 2 & 1 \\ 3 & 2 & 1 & 0 & 1 & 3 & 2 \\ 4 & 3 & 2 & 1 & 0 & 4 & 3 \\ 2 & 1 & 2 & 3 & 4 & 0 & 3 \\ 3 & 2 & 1 & 2 & 3 & 3 & 0 \end{pmatrix};$$

$$\mathrm{RD}(T) = \begin{pmatrix} 0 & 1 & \dfrac{1}{2} & \dfrac{1}{3} & \dfrac{1}{4} & \dfrac{1}{2} & \dfrac{1}{3} \\[8pt] 1 & 0 & 1 & \dfrac{1}{2} & \dfrac{1}{3} & 1 & \dfrac{1}{2} \\[8pt] \dfrac{1}{2} & 1 & 0 & 1 & \dfrac{1}{2} & \dfrac{1}{2} & 1 \\[8pt] \dfrac{1}{3} & \dfrac{1}{2} & 1 & 0 & 1 & \dfrac{1}{3} & \dfrac{1}{2} \\[8pt] \dfrac{1}{4} & \dfrac{1}{3} & \dfrac{1}{2} & 1 & 0 & \dfrac{1}{4} & \dfrac{1}{3} \\[8pt] \dfrac{1}{2} & 1 & \dfrac{1}{2} & \dfrac{1}{3} & \dfrac{1}{4} & 0 & \dfrac{1}{3} \\[8pt] \dfrac{1}{3} & \dfrac{1}{2} & 1 & \dfrac{1}{2} & \dfrac{1}{3} & \dfrac{1}{3} & 0 \end{pmatrix}.$$

Therefore, the Harary index of the tree T is

$$H(T) = 12.$$

1.5 Modified Harary Index

There is a problem with the original definition of the Harary index. In this definition, the greater weights are given to the interior (inner) edges and smaller weights to the terminal (outer) edges of a (molecular) graph. This can be remedied by the modification of the Harary index via the Harary matrix. One possible way of modifying the Harary index is as follows. Let us use as an illustrative example a tree T representing the carbon skeleton of 2,3-dimethylpentane (see Fig. 1.5). One can associate the Harary matrix with the superimposed structure SS generated as shown in Fig. 1.6.

The Harary matrix corresponding to the superimposed structure SS is given by:

$$RD(SS) = \begin{pmatrix} 0 & 1.7847 & 0 & 0 & 0 & 0 & 0 \\ 1.7847 & 0 & 2.6806 & 0 & 0 & 1.7847 & 0 \\ 0 & 2.6806 & 0 & 2.3194 & 0 & 0 & 1.8333 \\ 0 & 0 & 2.3194 & 0 & 1.5972 & 0 & 0 \\ 0 & 0 & 0 & 1.5972 & 0 & 0 & 0 \\ 0 & 1.7847 & 0 & 0 & 0 & 0 & 0 \\ 0 & 0 & 1.8333 & 0 & 0 & 0 & 0 \end{pmatrix}.$$

In this form, the Harary matrix is a sparse matrix and the summation of the elements in the upper or lower triangle of the matrix produces the Harary index:

$$H(T) = \frac{1}{2} \sum_{i=1}^{n} \sum_{j=1}^{n} [RD(SS)_{ij}] = 12.$$

One can modify the Harary matrix in its sparse form by replacing each nonzero element of the matrix by its reciprocal. In this way, one obtains the modified Harary matrix $^{m}(RD)$:

$$^{m}(RD)(T) = \begin{pmatrix} 0 & 0.5603 & 0 & 0 & 0 & 0 & 0 \\ 0.5603 & 0 & 0.3731 & 0 & 0 & 0.5603 & 0 \\ 0 & 0.3731 & 0 & 0.4311 & 0 & 0 & 0.5455 \\ 0 & 0 & 0.4311 & 0 & 0.6261 & 0 & 0 \\ 0 & 0 & 0 & 0.6261 & 0 & 0 & 0 \\ 0 & 0.5603 & 0 & 0 & 0 & 0 & 0 \\ 0 & 0 & 0.5455 & 0 & 0 & 0 & 0 \end{pmatrix}.$$

The superimposed structure corresponding to a tree T depicting the carbon skeleton of 2,3-dimethylpentane is given in Fig. 1.7. The modified Harary index (^{m}H) of the 2,3-dimethylpentane is then given as the sum of the elements in the upper (or lower) matrix triangle: 3.0964.

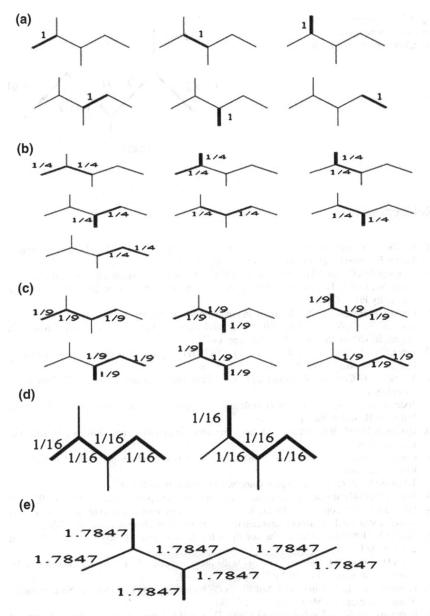

Fig. 1.6 Weights of individual edges making up paths of different lengths that are used to compute the modified Harary index of T. **a** Weighted paths of length 1; **b** Weighted paths of length 2; **c** Weighted paths of length 3; **d** Weighted paths of length 4 and **e** The superimposed structure SS

Fig. 1.7 The superimposed structure corresponding to the 2,3-dimethylpentane tree

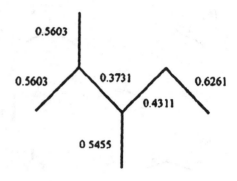

References

1. Bondy JA, Murty USR (1976) Graph theory with applications. Macmillan Press, New York
2. Harary F (1969) Graph theory. Addison-Wesley, Reading
3. Brouwer AE, Cohen AM, Neumaier A (1989) Distance-regular graphs. Springer, Berlin
4. Biggs NL (1993) Distance-transitive graphs. Algebraic graph theory, 2nd edn. Cambridge University Press, Cambridge
5. Buckley F, Harary F (1990) Distance in graphs. Addison-Wesley, Redwood
6. Goddard W, Oellermann OR (2011) Structural analysis of complex networks. Distance in graphs. Birkhäuser/Springer, New York, pp 49–72
7. Janežič D, Miličević A, Nikolić S, Trinajstić N (2007) Graph theoretical matrices in chemistry. University of Kragujevac, Kragujevac
8. Todeschini R, Consonni V (2000) Handbook of molecular descriptors. Wiley-VCH, Weinheim, pp 497–502
9. Todeschini R, Consonni V (2009) Molecular descriptors for chemoinformatics, vol I, vol II. Wiley-VCH, Weinheim, pp 934–938
10. Gutman I, Furtula B (eds) (2012) Distance in molecular graphs-theory. University of Kragujevac, Kragujevac
11. Gutman I, Furtula B (eds) (2012) Distance in molecular graphs-applications. University of Kragujevac, Kragujevac
12. Trinajstić N (1992) Chemical graph theory. CRC Press, Boca Raton
13. Wiener H (1947) Structural determination of paraffin boiling points. J Am Chem Soc 69:17–20
14. Hosoya H (1971) Topological index. A newly proposed quantity characterizing the topological nature of structural isomers of saturated hydrocarbons. Bull Chem Soc Jpn 44:2332–2339
15. Randić M (1993) Novel molecular descriptor for structure-property studies. Chem Phys Lett 211:478–483
16. Klein DJ, Lukovits I, Gutman I (1995) On the definition of the hyper-Wiener index for cycle-containing structures. J Chem Inf Comput Sci 35:50–52
17. Plavšić D, Nikolić S, Trinajstić N, Mihalić Z (1993) On the Harary index for the characterization of chemical graphs. J Math Chem 12:235–250
18. Ivanciuc O, Balaban TS, Balaban AT (1993) Design of topological indices. Part 4. Reciprocal distance matrix, related local vertex invariants and topological indices. J Math Chem 12:309–318
19. Mihalić Z, Trinajstić N (1992) A graph-theoretical approach to structure-property relationships. J Chem Educ 69:701–712
20. Xu K, Das KC (2011) On Harary index of graphs. Discret Appl Math 159:1631–1640
21. Zhou B, Du Z, Trinajstić N (2008) Harary index of landscape graphs. Int J Chem Model 1:35–44
22. Ivanciuc O (2000) QSAR comparative study of Wiener descriptors for weighted molecular graphs. J Chem Inf Comput Sci 40:1412–1422

23. Ivanciuc O, Ivanciuc T, Balaban AT (2000) The complementary distance matrix, a new molecular graph metric. ACH Models Chem 137:57–82
24. Gutman I, Furtula B, Petrović M (2009) Terminal Wiener index. J Math Chem 46:522–531
25. Székely LA, Wang H, Wu T (2011) The sum of distances between the leaves of a tree and the semi-regular property. Discret Math 311:1197–1203
26. Gutman I (1994) A formula for the Wiener number of trees and its extension to graphs containing cycles. Graph Theory Notes NY 27:9–15
27. Balaban AT (1982) Highly discriminating distance based topological index. Chem Phys Lett 89:399–404
28. Balaban AT (1983) Topological indices based on topological distances in molecular graphs. Pure Appl Chem 55:199–206
29. Dobrynin AA, Entringer R, Gutman I (2001) Wiener index of trees: theory and applications. Acta Appl Math 66:211–249
30. Dobrynin AA, Gutman I, Klavžar S, Žiget P (2002) Wiener index of hexagonal systems. Acta Appl Math 72:247–294
31. Feng L, Ilić A (2010) Zagreb, Harary and hyper-Wiener indices of graphs with a given matching number. Appl Math Lett 23:943–948
32. Furtula B, Gutman I, Tomović Ž, Vesel A, Pesek I (2002) Wiener-type topological indices of phenylenes. Indian J Chem A 41:1767–1772
33. Gutman I (1997) A property of the Wiener number and its modifications. Indian J Chem A 36:128–132
34. Gutman I, Rada J, Araujo O (2000) The Wiener index of starlike trees and a related partial order. MATCH Commun Math Comput Chem 42:145–154
35. Liu M, Liu B (2010) Trees with the seven smallest and fifteen greatest hyper-Wiener indices. MATCH Commun Math Comput Chem 63:151–170
36. Nikolić S, Trinajstić N, Mihalić Z (1995) The Wiener index: development and applications. Croat Chem Acta 68:105–129
37. Zhou B, Trinajstić N (2010) Mathematical properties of molecular descriptors based on distances. Croat Chem Acta 83:227–242
38. Bonchev D, Trinajstić N (1977) Information theory, distance matrix, and molecular branching. J Chem Phys 67:4517–4533
39. Mohar B, Babić D, Trinajstić N (1993) A novel definition of the Wiener index for trees. J Chem Inf Comput Sci 33:153–154
40. Gutman I (1994) Selected properties of the Schultz molecular topogical index. J Chem Inf Comput Sci 34:1087–1089
41. Nikolić S, Trinajstić N, Randić M (2001) Wiener index revisited. Chem Phys Lett 333:319–321
42. Zhou B (2010) Reverse Wiener index. In: Gutman I, Furtula B (eds) Novel molecular structure descriptors-theory and applications II. University of Kragujevac, Kragujevac, pp 193–204
43. Ghorbani M (2012) Computing Wiener index of chemical compounds by cut method. In: Gutman I, Furtula B (eds) Distance in molecular graphs-theory. University of Kragujevac, Kragujevac, pp 71–83
44. Ghorbani M, Ashrafi AR, Zousefi S (2012) Wiener index of nanotubes and nanotori. In: Gutman I, Furtula B (eds) Distance in molecular graphs-application. University of Kragujevac, Kragujevac, pp 157–166
45. Balaban TS, Filip PA, Ivanciuc O (1992) Computer generation of acyclic graphs based on local vertex invariants and topological indices. Derived canonical labelling and coding trees and alkanes. J Math Chem 11:79–105
46. Lučić B, Sović I, Plavšić D, Trinajstić N (2012) Harary matrices: definitions, properties and applications. In: Gutman I, Furtula B (eds) Distance in molecular graphs-applications. University of Kragujevac, Kragujevac, pp 3–26

Chapter 2
Extremal Graphs with Respect to Harary Index

In recent years, characterizing the extremal (maximal or minimal) graphs in a given set of graphs with respect to some distance-based topological index has become an important direction in chemical graph theory.

Let us briefly recall the chemical background of this problem as follows. A class of molecular graphs representing carbon compounds is a class of connected graphs with maximum degree at most 4. It models the skeletons of hydrocarbons [1], an important class of molecules in organic chemistry. The bounds of a molecular descriptor are important information of a molecular graph in the sense that they establish the approximate range of the descriptor in terms of molecular structural parameters. Therefore, it is important to establish the (lower or upper) bounds for topological indices and to characterize the corresponding extremal graphs at which the lower or upper bounds are attained.

Alternatively, topological index of a graph can be viewed as a graph invariant under the isomorphism of graphs, that is, for some topological index TI, $\mathrm{TI}(G) = \mathrm{TI}(H)$ if $G \cong H$. Therefore, the results in this chapter can also be seen as a topic in extremal graph theory. For some other interesting results in extremal graph theory, see [2].

In this chapter, we determine the upper or lower bounds on the Harary indices of graphs in various sets of structures, including general graphs, trees and generalized trees, and characterize the corresponding extremal graphs at which these bounds are attained. For some recent related results to this topic, see a recent survey [3].

2.1 General Graphs

In this section, we present some extremal results on general graphs with respect to Harary index.

Denote by $G^{\circledast} = (V, E)$ a graph with diameter d ($3 \le d \le 4$ and $|V(G^{\circledast})| \ge d+2$) such that, for any two distinct vertices $u \in V(G^{\circledast}) \setminus V(P_{d+1})$ and $v \in V(G^{\circledast})$, $d_{G^{\circledast}}(u, v) = 1 \ or \ 2$ where P_{d+1} is a path with $d + 1$ vertices in $G^{\circledast}$. Two graphs depicted in Fig. 2.1 are all of $G^{\circledast}$ type.

© The Author(s) 2015
K. Xu et al., *The Harary Index of a Graph*,
SpringerBriefs in Mathematical Methods, DOI 10.1007/978-3-662-45843-3_2

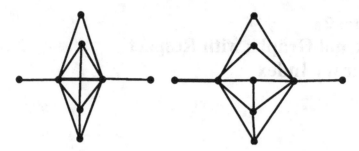

Fig. 2.1 Examples of graphs of $G^{\circledast}$-type

Theorem 2.1.1 ([4]) *Let G be a connected graph of order n and with m edges and diameter $D(G) = d$. Set $A = H(P_{d+1})$. Then*

$$A + \frac{n(n-1) + 2(m-d)(d-1)}{2d} - \frac{d+1}{2} \le H(G)$$

$$\le A + \frac{n(n-1) + 2m}{4} - \frac{d(d+3)}{4}$$

with left equality holding if and only if G is a graph with diameter $d \le 2$ or $G \cong P_n$, and right equality holding if and only if G is a graph with diameter $d \le 2$ or $G \cong P_n$, or G is isomorphic to some $G^{\circledast}$.

A connected graph G is called a *cactus* if each block of G is either an edge or a cycle. Denote by C at (n, r) the set of connected cacti possessing n vertices and r cycles. Let $C^0(n, r)$ be the cactus graph obtained from a star S_n by adding r independent edges between the leaves of S_n.

Theorem 2.1.2 ([5, 6]) *Let G be any graph in C at (n, r). Then we have*

$$H(G) \le \frac{1}{4}(n - 2r - 1)(n - 2r - 2) - r^2 + (n - 1)(r + 1)$$

with equality holding if and only if $G \cong C^0(n, r)$.

Theorem 2.1.3 ([7]) *Among all cacti of order $2n$ and with a perfect matching, the graph $C^0(2n, n - 1)$ is the unique graph having the maximal Harary index.*

Let $C^*_{n,k}$ be a cactus obtained by identifying the vertex of degree $n - 4$ of C^0 $(n - 3, \frac{n-k-4}{2})$ with one vertex of C_4. For example, the graph $C^*_{11,3}$ is shown in Fig. 2.2.

Theorem 2.1.4 ([7]) *Among all cacti of order n and with k cut edges, the graph $C^0(n, \frac{n-k-1}{2})$ is the unique graph with maximal Harary index when $n - k$ is odd; and $C^*_{n,k}$ uniquely has the maximal Harary index if $n - k$ is even.*

Fig. 2.2 The cactus $C^*_{11,3}$

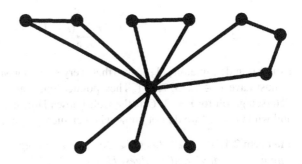

Theorem 2.1.5 ([7]) *Among all cacti of order n and with k pendant vertices, the graph $C^0(n, \frac{n-k-1}{2})$ is the unique graph with maximal Harary index when $n - k$ is odd; and $C^*_{n,k}$ uniquely has the maximal Harary index if $n - k$ is even.*

Denote by $C^\dagger(2n, r)$ a graph of order $2n$ obtained by attaching $n - r - 1$ paths of length 2 at the vertex of maximum degree in $C^0(2r + 2, r)$.

Theorem 2.1.6 ([6]) *Let $G \in C$ at $(2n, r)$ with a perfect matching. Then we have*

$$H(G) \le \frac{1}{24}(n - r - 1)(23n + 17r - 2) + 2n + r^2 - 1$$

with equality holding if and only if $G \cong C^\dagger(2n, r)$.

Let $1 \le k < n$ and KC^k_n be the graph obtained by attaching k pendant vertices to one vertex of the complete graph K_{n-k}.

Theorem 2.1.7 ([8]) *Among all connected graphs with n vertices and k cut edges, the graph KC^k_n uniquely has the maximal Harary index.*

Note that the *kite graph* $Ki_{n,k}$ is obtained by identifying one vertex of K_k with one pendant vertex of P_{n-k+1} and the *Turán graph* $T_n(k)$ is a complete k-partite graph of order n in which any two partition sets differ in size by at most one.

Theorem 2.1.8 ([9]) *Among all connected graphs with n vertices and clique number k, the Turán graph $T_n(k)$ uniquely has the maximal Harary index, the kite graph $Ki_{n,k}$ uniquely has the minimal Harary index.*

Moreover, in [10], the authors also determined some extremal bipartite graphs with respect to Harary index, which are all complete bipartite graphs. Hence, these results can be viewed as the special cases of Theorem 2.1.8.

Theorem 2.1.9 ([9]) *Among all connected graphs with n vertices and chromatic number k, the Turán graph $T_n(k)$ uniquely has the maximal Harary index, the kite graph $Ki_{n,k}$ uniquely has the minimal Harary index.*

In graph theory, the well-known *Moore graph* is a r-regular graph with diameter k whose order attains the upper bound

$$1 + r \sum_{i=0}^{k-1} (r-1)^i.$$

Hoffman and Singleton [11] proved that every r-regular Moore graph G with diameter 2 must have $r \in \{2, 3, 7, 57\}$. They pointed out that $G \cong C_5$ if $r = 2$, G is just Petersen graph for $r = 3$; G is the well-known Hoffman-Singleton graph for $r = 7$ and while $r = 57$ we do not know whether such graph G exists or not.

Theorem 2.1.10 ([12]) *Let G be a connected triangle- and quadrangle-free graph with $n \geq 2$ vertices and m edges. Then*

$$H(G) \leq \frac{n(n-1)}{4} + \frac{m}{2}$$

with equality holding if and only if G is a star or a Moore graph of diameter 2.

In the next theorem, we will characterize the extremal graphs maximizing the Harary index among all connected graphs with a given matching number. Obviously, either $G = C_3$ or $G = S_n$ holds for any connected graph G with $n \geq 2$ vertices and matching number $\beta = 1$. For the connected graph G with $n \geq 4$ vertices and matching number $\beta \geq 2$, we have

Theorem 2.1.11 ([13]) *Let G be a connected graph with $n \geq 4$ vertices and matching number β, where $2 \leq \beta \leq \left\lfloor \frac{n}{2} \right\rfloor$.*

(1) *If $\beta = \left\lfloor \frac{n}{2} \right\rfloor$, then $H(G) \leq H(K_n)$ with equality holding if and only if $G \cong K_n$;*

(2) *If $\frac{2n}{5} < \beta \leq \left\lfloor \frac{n}{2} \right\rfloor - 1$, then $H(G) \leq H(K_1 \bigvee (K_{2\beta-1} \cup \overline{K_{n-2\beta}}))$ with equality holding if and only if $G \cong K_1 \bigvee (K_{2\beta-1} \cup \overline{K_{n-2\beta}})$;*

(3) *If $2 \leq \beta < \frac{2n}{5}$, then $H(G) \leq H(K_\beta \bigvee \overline{K_{n-\beta}})$ with equality holding if and only if $G \cong K_\beta \bigvee \overline{K_{n-\beta}}$;*

(4) *If $\beta = \frac{2n}{5}$, then $H(G) \leq H(K_\beta \vee \overline{K_{n-\beta}}) = H(K_1 \vee (K_{2\beta-1} \cup \overline{K_{n-2\beta}}))$ with equality holding if and only if $G \cong K_\beta \bigvee \overline{K_{n-\beta}}$ or $G \cong K_1 \bigvee (K_{2\beta-1} \cup \overline{K_{n-2\beta}})$.*

By the definition of Harary index, one can easily observe that any edge addition will increase the Harary index. Thus, we have

Proposition 2.1.12 ([9]) *Let G be a connected graph with $e \notin E(G)$. Then we have $H(G) < H(G + e)$.*

By Proposition 2.1.12, it easily follows that

Theorem 2.1.13 ([12]) *Let G be a connected graph of order n. Then $H(G) \leq H(K_n)$ with equality holding if and only if $G \cong K_n$.*

A graph G is called *quasi-tree graph* if there exists a vertex $x \in V(G)$ such that $G - x$ is a tree. Clearly, any tree is a quasi-tree graph since the deletion of any pendant vertex will deduce another new tree. So, we call any tree a trivial quasi-tree graph, and other quasi-tree graphs are called nontrivial quasi-tree graphs. Very recently in [14] we introduced a new definition of k-generalized quasi-tree graph. A graph G is called *k-generalized quasi-tree graph* if there exists a subset $V_k \subseteq V(G)$ with $|V_k| = k$ such that $G - V_k$ is a tree but, for any subset $V_{k-1} \subseteq V(G)$ with cardinality $k - 1$, $G - V_{k-1}$ is not a tree. For $k \geq 2$, we denote by $\mathcal{QT}^{(k)}(n)$ the set of k-generalized quasi-tree graphs of order n. Here, we think nontrivial quasi-tree graphs and generalized quasi-tree graphs as general graphs because of their more complicated structure [14] than unicyclic or bicyclic graphs.

Let $C_k((n-k)^1)$ be a graph obtained by attaching a path of length $n-k$ to any one vertex of C_k. We denote by $C_{3,3}^{n-5}$ (see Fig. 2.3) a graph obtained by connecting two vertex-disjoint triangles by a path of length $n - 5$. Extremal graphs with respect to Harary index are characterized, respectively, in the following three theorems among all nontrivial quasi-tree graphs of order $n \geq 4$ and k-generalized graphs of order $n \geq 6$ (including the minimal case for $k = 2$ and the maximal case for all values of k).

Theorem 2.1.14 ([14]) *Let G be a nontrivial quasi-tree graph of order $n \geq 4$. Then we have*

$$3 + n \sum_{k=2}^{n-2} \frac{1}{k} \leq H(G) \leq \frac{(n-2)(n+5)}{4} + 1$$

with left equality holding if and only if $G \cong C_3((n-3)^1)$, and right equality holding if and only if $G \cong K_2 \bigvee \overline{K_{n-2}}$.

Theorem 2.1.15 ([14]) *Let G be a 2-generalized quasi-tree graph of order $n \geq 6$. Then we have*

$$H(G) \geq 5 + n \sum_{k=2}^{n-3} \frac{1}{k} + \frac{1}{n-3}$$

with equality holding if and only if $G \cong C_{3,3}^{n-5}$.

Fig. 2.3 The graph $C_{3,3}^{n-5}$

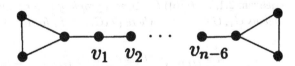

Theorem 2.1.16 ([14]) *For any graph $G \in QT^{(k)}(n)$ with $n \geq 6$, we have*

$$H(G) \leq \frac{n(n-1)}{4} + \frac{(k+1)(n-k-1)}{2} + \frac{(k+1)k}{4}$$

with equality holding if and only if $G \cong K_{k+1} \bigvee \overline{K_{n-k-1}}$.

For a connected graph G, the kth power G^k (see Ref. [15]) is a new graph with vertex set $V(G^k) = V(G)$ such that two vertices are adjacent in G^k if and only if they are at distance at most k in G. The bounds on Harary index have been presented in the following theorem among all kth power of trees. Moreover the corresponding extremal graphs were also characterized implicitly with respect to Harary index.

Theorem 2.1.17 ([15]) *For any tree T of order n, we have*

$$H(P_n^k) \leq H(T^k) \leq H(S_n^k)$$

with left equality holding if and only if $T^k \cong P_n^k$ and right equality holding if and only if $T^k \cong S_n^k$.

From Proposition 2.1.12, the corollary below can be easily obtained.

Corollary 2.1.18 ([15]) *Let G be a connected graph of order n. Then we have $H(P_n^k) \leq H(G^k)$.*

In the several theorems below, we present some extremal results with respect to Harary index on disconnected graphs. First, we define

$$f(n,k) = \begin{cases} k + n \sum_{l=2}^{r-1} \dfrac{1}{l} + \dfrac{s(r+1)}{r} & \text{if } k \leq \dfrac{n}{2}; \\[2ex] n - k & \text{if } k > \dfrac{n}{2}, \end{cases}$$

where r, s are integers with $n = rk + s$ and $0 \leq s < k$.

Theorem 2.1.19 ([16]) *Let G be a graph of order n and with k components where $1 \leq k \leq n$. Then we have*

$$f(n,k) \leq H(G) \leq \frac{(n-k+1)(n-k)}{2}$$

with left equality holding if and only if $G \cong (k-s)P_r \cup sP_{r+1}$ and right equality holding if and only if $G \cong (k-1)K_1 \cup K_{n-k+1}$.

Theorem 2.1.20 ([16]) *Let G be a graph of order n and with m edges and k components $G_1, G_2, \ldots, G_k$ where $|V(G_i)| = n_i$ for $i = 1, 2, \ldots, k$. Then we have*

$$\sum_{i=1}^{k} H(P_{n_i}) + \frac{m-n+k}{2} \leq H(G) \leq \frac{\sum_{i=1}^{k} n_i^2}{4} - \frac{n}{4} + \frac{m}{2}$$

with left equality holding if and only if $G_i \cong P_{n_i}$ or K_3 for $i = 1, 2, \ldots, k$ and right equality holding if and only if G_i has diameter at most 2 for $i = 1, 2, \ldots, k$.

Theorem 2.1.21 ([16]) *Let G be a graph of order n and with m edges and k components where $1 \leq k \leq n$. Then we have*

$$f(n, k) + \frac{m - n + k}{2} \leq H(G) \leq \frac{(n - k + 1)(n - k)}{4} + \frac{m}{2}$$

with left equality holding if and only if $G \cong (k - s)P_r \cup s P_{r+1}$ and right equality holding if and only if $G \cong (k-1)K_1 \cup K_{n-k+1}$, where r, s are integers with $n = rk + s$ and $0 \leq s < k$.

2.2 Trees

When we study some property of graphs, a tree is generally viewed as the simplest graph to consider first as a starting point. In this section, we report some extremal results on trees with respect to Harary index.

Theorem 2.2.1 ([17, 18]) *Let T be a tree of order n. Then we have*

$$H(P_n) \leq H(T) \leq H(S_n)$$

with left equality holding if and only if $T \cong P_n$, and right equality holding if and only if $T \cong S_n$.

By Proposition 2.1.12, among all connected graphs, the extremal graph with the minimal Harary index must be a tree. Thus, by Theorem 2.2.1, we have

Corollary 2.2.2 ([12]) *Let G be a connected graph of order n. Then we have $H(G) \geq H(P_n)$ with equality holding if and only if $G \cong P_n$.*

Although by now, in chemical graph theory, the measure of branching cannot be formally defined [19], there are several properties that any proposed measure has to satisfy [20, 21]. Basically, a topological index (TI) acceptable as a measure of branching must satisfy the inequalities

$$\mathrm{TI}(S_n) < \mathrm{TI}(T) < \mathrm{TI}(P_n) \quad or \quad \mathrm{TI}(P_n) < \mathrm{TI}(T) < \mathrm{TI}(S_n)$$

for any tree T of order $n \geq 5$ different from S_n and P_n. From Theorem 2.2.1, we find that Harary index (H) satisfies the basic requirement to be a branching index.

Taking Theorem 2.2.1 into consideration, we naturally ask: *Which trees have the extremal Harary indices among the trees of order n different from S_n and P_n?* The next two theorems will give an answer to this question, in which the ordering of trees will be extended with respect to Harary index.

Before stating these two theorems, we first introduce some necessary notations and definitions. A vertex v of a tree T is called a *branching point* if $d_T(v) \geq 3$. Let $T_n(n_1, n_2, \ldots, n_m)$ be a starlike tree of order n obtained by inserting $n_1 - 1, \ldots, n_m - 1$ vertices into m edges of the star S_{m+1}, respectively, where $n_1 + \cdots + n_m = n - 1$. Note that any tree with exactly one branching point is a starlike tree. Assume that T is a tree of order n with exactly two branching points v_1 and v_2 with $d_T(v_1) = r$ and $d_T(v_2) = t$. The orders of $r - 1$ components, which are paths, of $T - v_1$ are $p_1, \ldots, p_{r-1}$, the order of the component which is not a path of $T - v_1$ is $p_r = n - p_1 - \cdots - p_{r-1} - 1$. The orders of $t - 1$ components, which are paths, of $T - v_2$ are $q_1, \ldots, q_{t-1}$, the order of the component which is not a path of $T - v_2$ is $q_t = n - q_1 - \cdots - q_{t-1} - 1$. We denote this tree by $T = T_n(p_1, \ldots, p_{r-1}; q_1, \ldots, q_{t-1})$, where $r \leq t$, $p_1 \geq \cdots \geq p_{r-1}$ and $q_1 \geq \cdots \geq q_{t-1}$.

For convenience, when considering the trees $T_n(n_1, n_2, \ldots, n_k, \ldots, n_m)$ or $T_n(p_1, \ldots, p_k, \ldots, p_{r-1}; q_1, \ldots, q_k, \ldots, q_{t-1})$, we use the symbols $n_k^{l_k}$ or $p_k^{l_k}$ (resp. $q_k^{l_k}$) to indicate that the number of n_k or p_k (resp. q_k) is $l_k > 1$ in the following. For example, $T_{16}(2, 2, 3, 3, 5)$ will be written as $T_{16}(2^2, 3^2, 5^1)$. Let $T_2, T_3, \ldots, T_8$ be the trees of order $n \geq 14$ as shown in Fig. 2.4.

Theorem 2.2.3 ([18]) *Suppose that T is a tree of order $n \geq 16$. Then we have*

$$H(P_n) < H(T_n(n-3, 1^2)) < H(T_n(n-4, 2, 1)) < H(T_n(1^2; 1^2))$$
$$< H(T_n(n-5, 3, 1)) < H(T_n(1^2; 2, 1)) < H(T_n(n-4, 1^3)) < H(T).$$

Theorem 2.2.4 ([18]) *Suppose that T is a tree of order $n \geq 16$. Then we have*

$$H(T) < H(T_8) < H(T_7) < H(T_6) < H(T_5)$$
$$< H(T_4) < H(T_3) < H(T_2) < H(S_n).$$

In the theorem below, we assume that $n - 1 = kq + r$ with $0 \leq r < k$, that is, $q = \left\lfloor \frac{n}{k} \right\rfloor$. Obviously, we have $n - 1 = k \left\lfloor \frac{n}{k} \right\rfloor + r = (k - r) \left\lfloor \frac{n}{k} \right\rfloor + r \left\lceil \frac{n}{k} \right\rceil$.

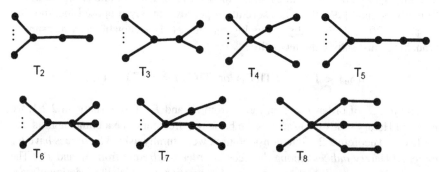

Fig. 2.4 The trees $T_2, T_3, \ldots, T_8$ encountered in Theorem 2.2.4

Theorem 2.2.5 ([22]) *Let T be a tree with n vertices and k pendant vertices, where $2 \leq k \leq n - 2$. Then*

$$H(T) \leq H\left(T_n\left(\left\lceil \frac{n}{k} \right\rceil^r, \left\lfloor \frac{n}{k} \right\rfloor^{k-r}\right)\right)$$

with equality holding if and only if $T \cong T_n(\left\lceil \frac{n}{k} \right\rceil^r, \left\lfloor \frac{n}{k} \right\rfloor^{k-r})$.

Recall that $T_n(2^{\beta-1}, 1^{n-2\beta+1})$ is a tree defined as above. Clearly, the matching number of $T_n(2^{\beta-1}, 1^{n-2\beta+1})$ is β, and there is exactly one tree with n vertices and matching number $\beta = 1$, which is just the star S_n. Recently, the maximal Harary index in the class of trees with n vertices and matching number $\beta \geq 2$ were determined in the following theorem.

Theorem 2.2.6 ([22, 23]) *Let T be a tree with n vertices and matching number $2 \leq \beta \leq \left\lfloor \frac{n}{2} \right\rfloor$. Then*

$$H(T) \leq H(T_n(2^{\beta-1}, 1^{n-2\beta+1}))$$

where the equality holds if and only if $T \cong T_n(2^{\beta-1}, 1^{n-2\beta+1})$.

It is well-known that $\alpha + \beta = n$ for a bipartite graph G of order n and with matching number β and independence number α (see, e.g., [24]). Therefore, the following corollary can be easily obtained from Theorem 2.2.6.

Corollary 2.2.7 ([22, 23]) *Let T be a tree with n vertices and independence number α. Then*

$$H(T) \leq H(T_n(2^{n-\alpha-1}, 1^{2\alpha-n+1}))$$

with equality holding if and only if $T \cong T_n(2^{n-\alpha-1}, 1^{2\alpha-n+1})$.

For $2 \leq \Delta \leq n - 1$, the *Volkmann tree* $V_{n,\Delta}$ is defined as follows [25, 26]:
If $n = \Delta + 1$, then $V_{n,\Delta}$ is just a star of order n;
For $n > \Delta + 1$, define n_i as $n_i = 1 + \sum_{j=1}^{i} \Delta(\Delta - 1)^j$ for $i = 1, 2, \ldots$, and choose k such that $n_{k-1} < n \leq n_k$.

Then calculate the parameters m and h by $m = \dfrac{n - n_{k-1}}{\Delta - 1}$ and $h = n - n_{k-1} - (\Delta - 1)m$.

The vertices of $V_{n,\Delta}$ are arranged into $k + 1$ levels. In level 0, there is only one vertex labeled as $v_{0,1}$. In level i for $i = 1, 2, \ldots, k - 1$, there are $\Delta(\Delta - 1)^i$ vertices labeled as $v_{i,1}, v_{i,2}, \ldots, v_{i,\Delta(\Delta-1)^i}$. These are connected (in that order) to the vertices in level i, so that $\Delta - 1$ vertices from level i are adjacent to each vertex from level $i - 1$. At level k there are $n - n_{k-1}$ vertices, labeled as $v_{k,1}, v_{k,2}, \ldots, v_{k,n-n_{k-1}}$. They are connected (in that order) to the vertices in level $k - 1$, so that $\Delta - 1$ vertices from level k are adjacent to vertices $v_{k-1,1}, v_{k-1,2}, \ldots, v_{k-1,m}$. The remaining h vertices at level k (if any) are connected to the vertex $v_{k-1,m+1}$ in level $k - 1$. To illustrate the structure of $V_{n,\Delta}$, we give an example in Fig. 2.5 for $n = 22$ and $\Delta = 4$.

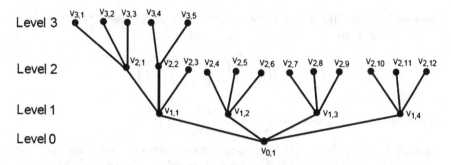

Fig. 2.5 The Volkmann tree $V_{22,4}$ with its vertices labeled

Theorem 2.2.8 ([15, 22, 27, 28]) *Let T be a tree with n vertices and maximum degree $\Delta \geq 3$. Then we have*

$$H(T_n(n - \Delta, 1^{\Delta-1})) \leq H(T) \leq H(V_{n,\Delta})$$

with left equality holding if and only if $T \cong T_n(n - \Delta, 1^{\Delta-1})$ and right equality holding if and only if $T \cong V_{n,\Delta}$.

In view of Proposition 2.1.12 and Theorem 2.2.8, the following result can be easily obtained.

Corollary 2.2.9 ([27]) *Let G be a connected graph of order n and with maximum degree Δ. Then we have*

$$H(G) \geq H(T_n(n - \Delta, 1^{\Delta-1}))$$

with equality holding if and only if $T \cong T_n(n - \Delta, 1^{\Delta-1})$.

In the next theorem, the extremal tree maximizing the Harary index is characterized completely among all trees of order n and with diameter d.

Theorem 2.2.10 ([22, 27]) *Let T be a tree with n vertices and diameter d, where $3 \leq d \leq n - 2$. Then*

$$H(T) \leq H\left(T_n\left(\left\lceil\frac{d}{2}\right\rceil, \left\lfloor\frac{d}{2}\right\rfloor, 1^{n-d-1}\right)\right)$$

with equality holding if and only if $T \cong T_n(\left\lceil\frac{d}{2}\right\rceil, \left\lfloor\frac{d}{2}\right\rfloor, 1^{n-d-1})$.

2.3 Generalized Trees

A unicyclic graph is a connected graph of order n and with n edges, which can be obtained by adding a new edge into a tree. Similarly, a bicyclic graph is a connected graph of order n and with $n + 1$ edges, which can be obtained by adding two new edges into a tree, i.e., by adding a new edge into a unicyclic graph. Therefore these two classes of graphs can be viewed as generalized trees. In this section we will determine some extremal results with respect to Harary index on these two classes of generalized trees.

Before presenting our main results, we first introduce some necessary notations. Denote by $C_k(n_1^{l_1}, n_2^{l_2}, \ldots, n_m^{l_m})$ the unicyclic graph obtained by attaching l_1 paths of length n_1, l_2 paths of length n_2, $\ldots$, l_m paths of length n_m, respectively, to one vertices of C_k, where $n_1 > n_2 > \cdots > n_m$. Note that the graph $C_k(l^1)$ defined in Sect. 2.1 is a special graph of $C_k(n_1^{l_1}, n_2^{l_2}, \ldots, n_m^{l_m})$. For example, the graph $C_5(4^1, 3^2, 2^2)$ is shown in Fig. 2.6. There are exactly two unicyclic graphs C_4 and $C_3(1^1)$ of order 4 with $H(C_4) = H(C_3(1^1))$. So we assume that $n \geq 5$ in the following theorem.

Theorem 2.3.1 ([29]) *Let G be a unicyclic graph of order $n \geq 5$. Then we have*

$$H(C_3((n-3)^1)) \leq H(G) \leq H(C_3(1^{n-3}))$$

where the left equality holds if and only if $G \cong C_3((n-3)^1)$, and the right equality holds if and only if $G \cong C_3(1^{n-3})$ for $n \geq 6$ and $G \cong C_3(1^{n-3})$ or $G \cong C_5$ for $n = 5$.

There is exactly one unicyclic graph C_3 with n vertices and matching number 1. For $n = 5$ and $\beta = 2$, we can easily check [29] that only two graphs C_n and $C_3(1^2)$ have the maximal Harary index among all unicyclic graphs of order n and with matching number 2. We find that [30] the unique graph $C_3(1^{n-3})$ has the maximal Harary index among these unicyclic graphs of order n and with matching number 2. Next, we present the extremal unicyclic graphs with maximal Harary index among all the unicyclic graphs with n vertices and matching number $\beta \geq 3$.

Theorem 2.3.2 ([31]) *Let G be a unicyclic graph with $n \geq 9$ vertices and matching number $\beta \geq 3$. Then*
$$H(G) \leq H(C_3(2^{\beta-2}, 1^{n-2\beta+1}))$$

with equality holding if and only if $G \cong C_3(2^{\beta-2}, 1^{n-2\beta+1})$.

Fig. 2.6 The graph
$C_5(4^1, 3^2, 2^2)$

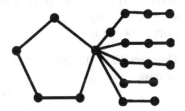

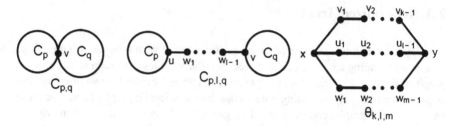

Fig. 2.7 The base graphs of type (I), (II), and (III)

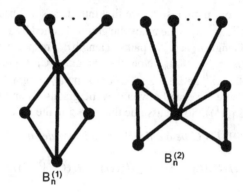

Fig. 2.8 The bicyclic graphs $B_n^{(1)}$ and $B_n^{(2)}$

In the next step, we turn to the determination of extremal Harary indices of bicyclic graphs. For any bicyclic graph G, the structure of cycles in G can be divided into the following three cases (see [32]):

(I)   The two cycles C_p and C_q in G have only one common vertex v;
(II) The two cycles C_p and C_q in G are linked by a path of length $l > 0$;
(III) The two cycles C_{l+k} and C_{l+m} in G have a common path of length $l > 0$.

As shown in Fig. 2.7, the bicyclic graphs $C_{p,q}$, $C_{p,l,q}$ and $\theta_{k,l,m}$ (where $1 \leq l \leq \min\{k, m\}$) corresponding to the above three cases are called the base subgraphs of bicyclic graph G of type (I), (II) and (III), respectively. For $n \geq 5$, let $B_n^{(1)}$ and $B_n^{(2)}$ be the bicyclic graphs as shown in Fig. 2.8.

For $n \geq 5$, let $B_n^{(0)}$ be a graph obtained by attaching a path of length $n - 4$ to one vertex of degree 2 pertaining to $\theta_{2,1,2}$ (see Fig. 2.9). In [33], the extremal graph with maximal Harary index has been determined among all bicyclic graphs of order n and with exactly two cycles. In the following two theorems, we characterize completely the extremal bicyclic graphs with respect to Harary index in the class of bicyclic graphs of order $n \geq 5$.

Theorem 2.3.3 ([29]) *Let G be a bicyclic graph of order $n \geq 5$ and $i \in \{1, 2\}$. Then we have*

$$H(G) \leq H(B_n^{(i)})$$

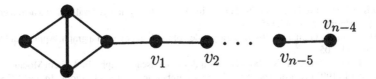

Fig. 2.9 The graph $B_n^{(0)}$

with equality holding if and only if $G \cong B_n^{(i)}$ for $n \geq 7$ and $G \cong B_n^{(i)}$ or $G \cong \theta_{2,2,3}$ for $n = 6$ and $G \cong B_n^{(i)}$ or $G \cong \theta_{2,1,3}$ or $G \cong K_{2,3}$ for $n = 5$.

Theorem 2.3.4 ([29]) *Let G be a bicyclic graph of order $n \geq 5$. Then we have*

$$H(B_n^{(0)}) \leq H(G)$$

where the equality holds if and only if $G \cong B_n^{(0)}$.

Although the extremal graphs with respect to Harary index have been completely determined among all unicyclic or bicyclic graphs of order n, there are still some interesting and challenging problems on this topic, such as determining the extremal graphs with respect to Harary index among all graph in some classes of unicyclic or bicyclic graphs, dealing with some extremal results among all connected graphs of order n and $m \geq n + 2$ edges. These problems seem to be attractive to many mathematical researchers.

References

1. Trinajstić N (1992) Chemical graph theory. CRC Press, Boca Raton
2. Bollobás B (2004) Extremal graph theory. Dover Publications, New York
3. Xu K, Liu M, Das KC, Gutman I, Furtula B (2014) A survey on graphs extremal with respect to distance-based topological indices. MATCH Commun Math Comput Chem 71:461–508
4. Das KC, Zhou B, Trinajstić N (2009) Bounds on Harary index. J Math Chem 46:1377–1393
5. Wang H, Kang L, On the Harary index of cacti. Util Math. In press
6. Zhu Z, Tao T, Yu J, Tan L (2014) On the Harary index of cacti. Filomat 28:493–507
7. Wang H, Kang L (2013) More on the Harary index of cacti. J Appl Math Comput 43:369–386
8. Xu K, Trinajstić N (2011) Hyper-Wiener and Harary indices of graphs with cut edges. Util Math 84:153–163
9. Xu K, Das KC (2011) On Harary index of graphs. Discret Appl Math 159:1631–1640
10. Cui Z, Liu B (2012) On Harary matrix, Harary index and Harary energy. MATCH Commun Math Comput Chem 68:815–823
11. Hoffman AJ, Singleton RR (1960) On Moore graphs with diameters 2 and 3. IBM J Res Dev 4:497–504
12. Zhou B, Cai X, Trinajstić N (2008) On Harary index. J Math Chem 44:611–618
13. Feng L, Ilić A (2010) Zagreb, Harary and hyper-Wiener indices of graphs with a given matching number. Appl Math Lett 23:943–948

14. Xu K, Wang J, Liu H (2014) The Harary index of ordinary and generalized quasi-tree graphs. J Appl Math Comput 45:365–374
15. Su G, Xiong L, Gutman I (2013) Harary index of the k-th power of a graph. Appl Anal Discret Math 7:94–105
16. Zhou B, Du Z, Trinajstić N (2008) Harary index of landscape graphs. Int J Chem Model 1:35–44
17. Gutman I (1997) A property of the Wiener number and its modifications. Indian J Chem A 36:128–132
18. Xu K (2012) Trees with the seven smallest and eight greatest Harary indices. Discret Appl Math 160:321–331
19. Bonchev D, Trinajstić N (1977) Information theory, distance matrix, and molecular branching. J Chem Phys 67:4517–4533
20. Fischermann M, Gutman I, Hoffmann A, Rautenbach D, Vidović D, Volkmann L (2002) Extremal chemical trees. Z Naturforsch 57a:49–52
21. Gutman I, Randić M (1977) Algebraic characterization of skeletal branching. Chem Phys Lett 47:15–19
22. Ilić A, Yu G, Feng L (2012) The Harary index of trees. Util Math 87:21–32
23. Das KC, Xu K, Gutman I (2013) On Zagreb and Harary indices. MATCH Commun Math Comput Chem 70:301–314
24. Bondy JA, Murty USR (1976) Graph theory with applications. Macmillan Press, New York
25. Fischermann M, Hoffmann A, Rautenbach D, Székely L, Volkmann L (2002) Wiener index versus maximum degree in trees. Discret Appl Math 122:127–137
26. Li W, Li X, Gutman I (2010) Volkmann trees and their molecular structure descriptors. In: Gutman I, Furtula B (eds) Novel molecular structure descriptors-theory and applications II. University of Kragujevac, Kragujevac, pp 231–246
27. He C, Chen P, Wu B (2010) The Harary index of a graph under perturbation. Discret Math Algorithms Appl 2:247–255
28. Wagner S, Wang H, Zhang X (2013) Distance-based graph invariants of trees and the Harary index. Filomat 27:39–48
29. Xu K, Das KC (2013) Extremal unicyclic and bicyclic graphs with respect to Harary index. Bull Malays Math Sci Soc 36:373–383
30. Diudea MV, Ivanciuc O, Nikolić S, Trinajstić N (1997) Matrices of reciprocal distance, polynomials and derived numbers. MATCH Commun Math Comput Chem 35:41–64
31. Xu K, Das KC, Hua H, Diudea MV (2013) Maximal Harary index of unicyclic graphs with given matching number. Stud Univ Babes-Bolyai Chem 58:71–86
32. Xu K, Gutman I (2011) The greatest Hosoya index of bicyclic graphs with given maximum degree. MATCH Commun Math Comput Chem 66:795–824
33. Yu G, Feng L (2010) On the maximal Harary index of a class of bicyclic graphs. Util Math 82:285–292

Chapter 3
Relation Between the Harary Index
and Related Topological Indices

In chemical graph theory, there are close connections among some topological indices of (molecular) graphs. In this chapter, we report several interesting relations between Harary index and other topological indices of graphs. Before doing it, we first introduce the definitions of some fundamental topological indices of graphs.

In 2000, Ivanciuc [1] and Ivanciuc et al. [2] first introduced the reciprocal complementary Wiener index (RCW(G)) of graph G as follows:

$$\text{RCW}(G) = \sum_{u,v \in V(G)} \frac{1}{d + 1 - d_G(u, v)},$$

where d is the diameter of graph G.

In chemical graph theory, two of the oldest topological indices are the well-known Zagreb indices first introduced in [3] where Gutman and Trinajstić examined the dependence of total π-electron energy on molecular structure and elaborated in [4]. For a (molecular) graph G, the first Zagreb index $M_1(G)$ and the second Zagreb index $M_2(G)$ are, respectively, defined as follows:

$$M_1 = M_1(G) = \sum_{v \in V(G)} d_G(v)^2, \quad M_2 = M_2(G) = \sum_{uv \in E(G)} d_G(u)d_G(v).$$

For some recent results of Zagreb indices and their multiplicative variants, please see [5–18].

3.1 Relation Between the Harary Index
and Reciprocal Wiener Index

In this section, we first compare the Harary index and reciprocal complementary Wiener index of graphs. Obviously, we have $H(S_n) \leq \text{RCW}(S_n)$ and $H(P_n) \geq \text{RCW}(P_n)$. But what is the answer for general graphs, even for trees? In the following,

© The Author(s) 2015
K. Xu et al., *The Harary Index of a Graph*,
SpringerBriefs in Mathematical Methods, DOI 10.1007/978-3-662-45843-3_3

Fig. 3.1 The double star
DS$_{5,4}$

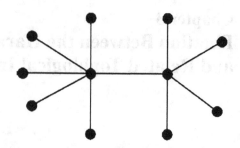

we turn to the comparison between H and RCW for some special trees. Denote by
DS$_{n_1,n_2}$ a double star obtained by adding a new edge between two central vertices
of stars S_{n_1+1} and S_{n_2+1}, respectively. For example, the double star DS$_{5,4}$ is shown
in Fig. 3.1.

Theorem 3.1.1 ([19]) *Let* DS$_{n_1,n_2}$ *be a double star with* $n_1 \geq 3$ *and* $n_2 \geq 2$. *Then*

$$H(\text{DS}_{n_1,n_2}) \leq \text{RCW}(\text{DS}_{n_1,n_2})$$

with equality holding if and only if $n_1 = 3$ *and* $n_2 = 2$.

For a graph G, we denote by G^* a new graph obtained by replacing each edge in
G with a path of length 2. As an example, the tree DS$_{5,4}^*$ is shown in Fig. 3.2.

Theorem 3.1.2 ([19]) *Let* DS$_{n_1,n_2}^*$ *be a tree with* $n_1 \geq 8n_2$. *Then*

$$\text{RCW}(\text{DS}_{n_1,n_2}^*) \leq H(\text{DS}_{n_1,n_2}^*).$$

In the two theorems below, we present the necessary conditions for $H(G) = $
RCW(G) for some special graphs.

Theorem 3.1.3 ([19]) *Let G be a graph of order n and with diameter 2. If $H(G) = $
RCW(G), then either n or $n - 1$ is divisible by 4.*

Two graphs satisfying the condition in Theorem 3.1.2 are shown in Fig. 3.3.

Fig. 3.2 The tree DS$_{5,4}^*$

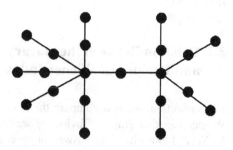

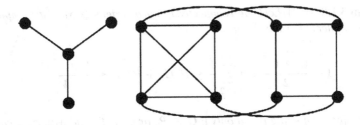

Fig. 3.3 Two graphs with $H(G) = RCW(G)$ and n is divisible by 4

Theorem 3.1.4 ([19]) *Let G be a triangle- and quadrangle-free graph of diameter 3. If $H(G) = RCW(G)$, then the first Zagreb index $M_1(G)$ is equal to the number of edges in $\overline{G}$.*

Recall that a graph G is self-complementary if G is isomorphic to its complement $\overline{G}$. The following corollary is obvious.

Corollary 3.1.5 *Let G be a self-complementary graph with diameter 2. Then we have $H(G) = RCW(G)$.*

In the following two theorems, we give some relations between Harary index $H(G)$ and Wiener index $W(G)$ of a graph G.

Theorem 3.1.6 ([20]) *Let G be a connected graph with $n \geq 2$ vertices, m edges, and diameter d. Then*

$$\left(W(G) - m - d\right)\left(H(G) - m - \frac{1}{d}\right) \geq \left(\frac{n(n-1)}{2} - m - 1\right)^2$$

with equality holding if and only if G has diameter of at most 2.

Theorem 3.1.7 ([21]) *Let $G \neq K_n$ be a connected graph of order n and with m edges and diameter d. Then we have*

$$m + \frac{\left(\frac{n(n-1)}{2} - m\right)^2}{W(G) - m} \leq H(G) \leq m + \frac{\left(\frac{n(n-1)}{2} - m\right)\left[2 + (\frac{n(n-1)}{2} - m - 1)(\frac{d}{2} + \frac{2}{d})\right]}{2\left(W(G) - m\right)}.$$

$$(3.1)$$

Each equality above holds if and only if G has diameter $d = 2$.

We denote by $I(G)$ a graph invariant of a graph G. The following relations on the respective invariants of G and its complement $\overline{G}$:

$$L_+(n) \leq I(G) + I(\overline{G}) \leq U_+(n), \quad L_\times(n) \leq I(G)I(\overline{G}) \leq U_\times(n)$$

are called Nordhaus–Gaddum type results with respect to the graph invariant $I(G)$ due to the work of Nordhaus and Gaddum [22]. In the next step, we present some results of Nordhaus–Gaddum type with respect to Harary index.

Theorem 3.1.8 ([23]) *Let G be a connected graph of order $n \geq 5$ and its complement $\overline{G}$ be the connected graph. Then*

$$1 + \frac{(n-1)^2}{2} + n \sum_{k=2}^{n-1} \frac{1}{k} \leq H(G) + H(\overline{G}) \leq \frac{3n(n-1)}{4}$$

with left equality holding if and only if $G \cong P_n$ or $G \cong \overline{P_n}$ and with right equality holding if and only if both G and $\overline{G}$ have diameter 2.

Theorem 3.1.9 ([19]) *Let G be a connected graph of order $n \geq 2$ with a connected complement $\overline{G}$. Denote by d and $\overline{d}$ the diameter of G and $\overline{G}$, respectively. Then*

$$H(G) + H(\overline{G}) \geq H(P_{k+1}) + \frac{n(n-1)}{2}\left(1 + \frac{1}{k}\right) - 3k + \frac{7}{2}$$

where $k = \max\{d, \overline{d}\}$. Moreover, the above equality holds if and only if both G and $\overline{G}$ have diameter 2.

Note that the lower bound in Theorem 3.1.9 is sometimes better than that given in Theorem 3.1.8. As an example, for the graph G_1 in Fig. 3.4, the lower bound for $H(G) + H(\overline{G})$ given in Theorem 3.1.9 is 31.5, better than that from Theorem 3.1.8, which is just 29.15.

Theorem 3.1.10 ([19]) *Let G be a connected graph of order $n \geq 2$ with a connected complement $\overline{G}$. If the graph G has diameter d, then*

$$H(G) + H(\overline{G}) \leq H(P_{d+1}) + \frac{3n(n-1)}{4} - \frac{d(d+3)}{4}$$

with equality holding if and only if both G and $\overline{G}$ have diameter 2, or $G \cong P_n$.

Because of the fact that $H(P_n) < \dfrac{(n-1)(n+2)}{4}$ when $n > 4$ (see Lemma 2.2 in [19]), the upper bound for $H(G) + H(\overline{G})$ in Theorem 3.1.10 is always better than that in Theorem 3.1.8.

Fig. 3.4 The graph G_1

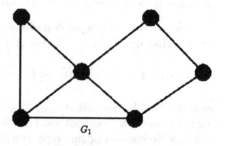

$$G_1$$

In the next theorem, the Nordhaus–Gaddum type result is presented for the kth power of connected graph G.

Theorem 3.1.11 ([24]) *Let G be a connected graph of order $n \geq 9$ and with a connected complement $\overline{G}$. Then we have*

$$\binom{n}{2} + \sum_{j=1}^{n-1} \frac{n-j}{\lceil \frac{j}{k} \rceil} = H(P_n^k) + H(\overline{P_n}^k) \leq H(G^k) + H(\overline{G}^k) \leq n\binom{n}{2}. \qquad (3.2)$$

Note that, for any graph G, at least one of G and $\overline{G}$ is connected. In the next theorem, we gave a Nordhaus–Gaddum type result for the case when exactly one of G or $\overline{G}$ is connected.

Theorem 3.1.12 ([25]) *Let G be a graph of order $n \geq 3$. If exactly one of G or $\overline{G}$ is connected, then we have*

$$\frac{n^2 - n}{2} \leq H(G) + H(\overline{G}) \leq \frac{3n^2 - 5n + 2}{4} \qquad (3.3)$$

with left equality holding if and only if G or $\overline{G}$ is nK_1, and right equality holding if and only if G or $\overline{G}$ is $K_1 \cup K_{n-1}$.

3.2 Relation Between the Harary Index and Zagreb Indices

In this section, we report some relations between Harary index and other topological indices of graphs, including the first and second Zagreb indices (M_1 and M_2), the largest eigenvalue of reciprocal distance matrix $RD(G)$ [26] of graph G.

Denote by G^{**} a triangle- and quadrangle-free graph of diameter 4 and order $n \geq 6$ such that, for any two distinct vertices $u, v \in V(G^{**})\backslash V(P_{d+1})$, $d_{G^{**}}(u, v) \leq 3$ where P_{d+1} is a path of order $d + 1$ in G^{**}. Two examples of graphs of G^{**}-type are shown in Fig. 3.5.

Theorem 3.2.1 ([19]) *Let G be a triangle- and quadrangle-free graph with $n \geq 2$ vertices, m edges, and diameter d. Then*

$$H(P_{d+1}) + \frac{d-2}{4d} M_1(G) + A_0 \leq H(G) \leq H(P_{d+1}) + \frac{1}{12} M_1(G) + A_1$$

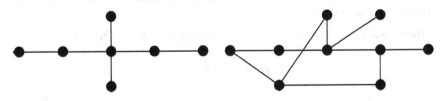

Fig. 3.5 Examples of graphs of the G^{**}-type

where $A_0 = \dfrac{n(n-1)-2}{2d} + \dfrac{m}{2} - 2(d-1)$ *and* $A_1 = \dfrac{n(n-1)+1}{6} + \dfrac{m}{2} - \dfrac{d^2}{6} - d.$

Moreover, the above left equality holds if and only if G is a graph of diameter at most 3 or a path P_n; the above right equality holds if and only if G is a graph of diameter at most 3 or a path P_n, or G is isomorphic to a graph G^{**}.

The following corollary can be easily deduced from Theorem 3.2.1.

Corollary 3.2.2 ([23]) *Let G be a triangle- and quadrangle-free graph with $n \geq 2$ vertices and m edges. Then*

$$H(G) \leq \frac{1}{12}M_1(G) + \frac{n(n-1)}{6} + \frac{m}{2}$$

with equality holding if and only if G is a graph of diameter at most 3.

In the following theorem, a relation is presented between $H(G) + H(\overline{G})$ and the first Zagreb index $M_1(G)$.

Theorem 3.2.3 ([19]) *Let G be a triangle- and quadrangle-free graph with $n \geq 2$ vertices, m edges, and with a connected complement $\overline{G}$. Then*

$$H(G) + H(\overline{G}) \leq \frac{1}{6}M_1(G) + \frac{7n(n-1)}{12} + \frac{n(n-1)^2}{12} - \frac{m(n-1)}{3}$$

with equality holding if and only if both G and $\overline{G}$ have diameter of at most 3.

Recall that, for a graph G of order n, the distance matrix $D(G)$ of G is an $n \times n$ matrix (d_{uv}) where d_{uv} is the distance between the vertices u and v in G [26], and the reciprocal distance matrix RD(G) of G is an $n \times n$ matrix (RD_{uv}) such that [26]

$$\mathrm{RD}_{uv} = \begin{cases} \dfrac{1}{d_{uv}} & \text{if } u \neq v; \\ 0 & \text{otherwise.} \end{cases}$$

Therefore, we have [27, 28] for a graph G, the Harary index of G can be also defined as follows:

$$H(G) = \sum_{u,v \in V(G)} \mathrm{RD}_{uv}.$$

Denote by $\lambda(G)$ the maximum eigenvalue of reciprocal distance matrix RD(G) of graph G. In the next theorem, an upper bound on $H(G)$ in terms of $\lambda(G)$ has been given for a connected graph G.

Theorem 3.2.4 ([29]) *Let G be a connected graph of order n. Then we have*

$$H(G) \leq \frac{n}{2}\lambda(G)$$

with equality holding if and only if RD(G) has equal row sums.

In the next theorem, a relation has been established between Harary index and two kinds of Zagreb indices (M_1 and M_2) for trees.

Theorem 3.2.5 ([30]) *Let T be a tree of order n and with diameter d. Then we have*

$$\left(\frac{1}{3} - \frac{1}{d}\right)M_2(T) + \left(\frac{1}{2d} - \frac{1}{12}\right)M_1(T) + A_0 \leq H(T) \leq \frac{1}{12}M_2(T) + \frac{1}{24}M_1(T) + A_1$$

where $A_0 = \dfrac{n^2}{2d} + \left(\dfrac{5}{6} - \dfrac{3}{2d}\right)n + \dfrac{1}{d} - \dfrac{5}{6}$ *and* $A_1 = \dfrac{1}{8}n^2 + \dfrac{11}{24}n - \dfrac{7}{12}$. *Moreover, both the above equalities hold if and only if T is a tree with diameter of at most 4.*

Although we have listed several results involving the relation between Harary index and related topological indices of a graph, we expect many interesting results on this topic which will be obtained by mathematical or chemical researchers in the near future.

References

1. Ivanciuc O (2000) QSAR comparative study of Wiener descriptors for weighted molecular graphs. J Chem Inf Comput Sci 40:1412–1422
2. Ivanciuc O, Ivanciuc T, Balaban AT (2000) The complementary distance matrix, a new molecular graph metric. ACH Models Chem 137:57–82
3. Gutman I, Trinajstić N (1972) Graph theory and molecular orbitals. III. Total π-electron energy of alternant hydrocarbons. Chem Phys Lett 17:535–538
4. Gutman I, Ruščić B, Trinajstić N, Wilcox CF (1975) Graph theory and molecular orbitals. XII. Acyclic polyenes. J Chem Phys 62:3399–3405
5. Das KC (2004) Maximizing the sum of the squares of degrees of a graph. Discret Math 257:57–66
6. Das KC, Gutman I (2004) Some properties of the second Zagreb index. MATCH Commun Math Comput Chem 52:103–112
7. Das KC, Gutman I, Zhou B (2009) New upper bounds on Zagreb indices. J Math Chem 46:514–521
8. Gutman I, Das KC (2004) The first Zageb index 30 years after. MATCH Commun Math Comput Chem 50:81–92
9. Gutman I, Furtula B, Toropov AA, Toropova AP (2005) The graph of atomic orbitals and its basic properties. 2. Zagreb Indices. MATCH Commun Math Comput Chem 53:225–230
10. Hosamani SM, Gutman I (2014) Zagreb indices of transformation graphs and total transformation graphs. Appl Math Comput 247:1156–1160
11. Miličević A, Nikolić S, Trinajstić N (2004) On reformulated Zagreb indices. Mol Divers 8:393–399
12. Nikolić S, Kovačević G, Miličević A, Trinajstić N (2003) The Zagreb indices 30 years after. Croat Chem Acta 76:113–124
13. Peng XL, Fang KT, Hu QN, Liang YZ (2004) Impersonality of the connectivity index and recomposition of topological indices according to different properties. Molecules 9:1089–1099
14. Xu K (2011) The Zagreb indices of graphs with a given clique number. Appl Math Lett 24:1026–1030
15. Xu K, Das KC (2012) Trees, unicyclic, and bicyclic graphs extremal with respect to multiplicative sum Zagreb index. MATCH Commun Math Comput Chem 68:257–272

16. Xu K, Das KC, Balachandran S (2014) Maximizing the Zagreb indices of (n, m)-graphs. MATCH Commun Math Comput Chem 72:641–654
17. Xu K, Hua H (2012) A unified approach to extremal multiplicative Zagreb indices for trees, unicyclic and bicyclic graphs. MATCH Commun Math Comput Chem 68:241–256
18. Zhou B (2004) Zagreb indices. MATCH Commun Math Comput Chem 52:113–118
19. Das KC, Zhou B, Trinajstić N (2009) Bounds on Harary index. J Math Chem 46:1377–1393
20. Xu K, Das KC (2011) On Harary index of graphs. Discret Appl Math 159:1631–1640
21. Das KC, Xu K, Cangul IN, Cevik AS, Graovac A (2013) On the Harary index of graph operations. J Ineq Appl 2013:1–16
22. Nordhaus EA, Gaddum JW (1956) On complementary graphs. Am Math Mon 63:175–177
23. Zhou B, Cai X, Trinajstić N (2008) On Harary index. J Math Chem 44:611–618
24. Su G, Xiong L, Gutman I (2013) Harary index of the kth power of a graph. Appl Anal Discret Math 7:94–105
25. Zhou B, Trinajstić N (2008) Harary index of landscape graphs. Int J Chem Model 1:35–44
26. Janežič D, Miličević A, Nikolić S, Trinajstić N (2007) Graph theoretical matrices in chemistry. University of Kragujevac, Kragujevac
27. Diudea MV (1997) Indices of reciprocal properties or Harary indices. J Chem Inf Comput Sci 37:292–299
28. Lučić B, Sović I, Plavšić D, Trinajstić N (2012) Harary matrices: definitions, properties and applications. In: Gutman I, Furtula B (eds) Distance in molecular graphs-applications. University of Kragujevac, Kragujevac, pp 3–26
29. Zhou B, Trinajstić N (2008) Maximum eigenvalues of the reciprocal distance matrix and the reverse Wiener matrix. Int J Quantum Chem 108:858–864
30. Das KC, Xu K, Gutman I (2013) On Zagreb and Harary indices. MATCH Commun Math Comput Chem 70:301–314

Chapter 4
Some Properties and Applications of Harary Index

In this chapter, we report on some properties and applications of the Harary index of a graph. Some mathematical properties of the Harary index are summarized in Sect. 4.1. Some of its applications in pure graph theory, mathematical chemistry, and structure–property modeling are reviewed in Sects. 4.2, 4.3, and 4.4, respectively.

For two graphs G_1 and G_2, there are many product graphs by G_1 and G_2. In general, these product graphs have a more complicated structure than G_1 and G_2. What is more, it is also difficult to compute the Harary index of product graphs, even though the tool (1.2) sometimes is much effective. In the first section, we investigate the effect on Harary index of a graph by inserting a new edge and present some formulae for computing the Harary indices of various product graphs.

A cycle in a graph G is called *Hamiltonian cycle* if it passes through all vertices in G exactly once. A graph G is called *Hamiltonian graph* or said to be *Hamiltonian* if it contains a Hamiltonian cycle. Similarly, a path in a graph G is known as *Hamiltonian path* if it passes though all vertices in G. A graph G is called *traceable* if it possesses a Hamiltonian path. Nowadays, some necessary or sufficient conditions are known to determine whether a graph G is Hamiltonian or not, including Dirac's condition [1], Ore's condition [2], Fan's condition [3], etc. Determining such a cycle or path exists in a graph is NP-complete. Finding a necessary and sufficient condition for a Hamiltonian graph has long been a fundamental but still open problem in pure graph theory. A nice survey on this topic is Ref. [4]. In the second section, we report two sufficient conditions for a graph to be traceable or Hamiltonian in terms of Harary index.

4.1 Some Properties of Harary Index

From the definition of Harary index, we can find that $H(G+uv) > H(G)$ for any two nonadjacent vertices $u, v \in V(G)$. But how large is the value of $H(G+uv) - H(G)$ for any two nonadjacent vertices $u, v \in V(G)$? In the next two theorems, we give the range of $H(G + uv) - H(G)$ for any two nonadjacent vertices $u, v \in V(G)$.

© The Author(s) 2015
K. Xu et al., *The Harary Index of a Graph*,
SpringerBriefs in Mathematical Methods, DOI 10.1007/978-3-662-45843-3_4

Theorem 4.1.1 ([5]) *Let G be a connected graph with $n \geq 2$ vertices, m edges, and diameter d. If there exist two nonadjacent two vertices $u, v \in V(G)$, then*

$$\frac{1}{2} \leq H(G + uv) - H(G) \leq 1 - \frac{1}{d} + \frac{n(n-1) - 2m - 2}{2}\left(\frac{1}{2} - \frac{1}{d}\right)$$

with left equality holding if and only if $d_G(u) = d_G(v) = 1$ and $d_G(u, v) = 2$, and right equality holding if and only if G has diameter 2.

Theorem 4.1.2 ([5]) *Let G be a triangle- and quadrangle-free graph with $n \geq 2$ vertices, m edges, and diameter d. If there exist two nonadjacent vertices $u, v \in V(G)$, then*

$$\frac{1}{2} \leq H(G + uv) - H(G) \leq 1 - \frac{1}{d} + \frac{n(n-1) - M_1(G) - 2}{2}\left(\frac{1}{2} - \frac{1}{d}\right)$$

with left equality holding if and only if $d_G(u) = d_G(v) = 1$ and $d_G(u, v) = 2$, and right equality holding if and only if G has diameter 2.

In graph theory, a graph product $G * H$ of graphs G and H is a new graph on the vertex set $V(G) \times V(H)$, while its edges are determined by a function of the edges of the factors (see [6]). For details on the properties and applications of product graphs, the readers are referred to the Handbook [7]. In what follows we will deal with the Harary index of various graph products.

Before stating our main results on the formulae of Harary index for graph products, we first recall the definitions of some graph products. For two graphs, G_1 of order n_1 and G_2 of order n_2, we introduce some definitions of graph products by G_1 and G_2. The join $G_1 \bigvee G_2$ is the same as defined in Sect. 1.1. The *corona product* $G_1 \circ G_2$ is a new graph obtained by taking one copy of G_1 and n_1 copies of G_2 and joining the ith vertex of G_1 to every vertex in ith copy of G_2 where $i = 1, 2, \ldots, n_1$. Clearly, $G_1 \circ G_2$ is connected when G_1 is connected. In general, $G_1 \circ G_2$ is not isomorphic to $G_2 \circ G_1$. The *Cartesian product* $G_1 \times G_2$ is a new graph with vertex set $V(G_1) \times V(G_2)$ and the vertices (u, v) and (x, y) are adjacent when $u = x$ and $vy \in E(G_2)$, or $ux \in E(G_1)$ and $v = y$. The *lexicographic product* $G_1[G_2]$ is a new graph with vertex set $V(G_1) \times V(G_2)$ and the vertices (u, v) and (x, y) are adjacent when $ux \in E(G_1)$ or $u = x$ and $vy \in E(G_2)$. The *disjunction* $G_1 \otimes G_2$ is a new graph with vertex set $V(G_1) \times V(G_2)$ and the vertices (u, v) and (x, y) are adjacent whenever $ux \in E(G_1)$ or $vy \in E(G_2)$. We define

$$H_t(G) = \sum_{u,v \in V(G)} \frac{1}{d_G(u, v) + t}$$

where the summation goes over all unordered pairs of vertices in graph G. Moreover, $H_t(G)$ can be viewed as a variant of Harary index (for it, we choose $t = 0$).

Theorem 4.1.3 ([8]) *Let G_i be a graph of order n_i and with m_i edges for $i = 1, 2$. Then we have*

$$H\left(G_1 \bigvee G_2\right) = \frac{1}{2}(n_1 + n_2 + m_1 m_2) + \frac{1}{4}(n_1 + n_2)(n_1 + n_2 - 1).$$

Theorem 4.1.4 ([8]) *Let G_i be a graph of order n_i and with m_i edges for $i = 1, 2$. Then we have*

$$H(G_1 \circ G_2) = H(G_1) + n_2 H_1(G_1) + n_2^2 H_2(G_1) + \frac{1}{4}(n_2 + 3)n_1 n_2 + \frac{1}{2}n_1 m_2.$$

Given a graph G with vertex set V of cardinality n, thorny graph $G(p_1, p_2, \ldots, p_n)$ is first introduced by Gutman [9], which is a graph obtained by attaching p_i pendant vertices to the ith vertex for $i = 1, 2, \ldots, n$. In particular, if $p_1 = p_2 = \cdots = p_n = p$, we denote by $G^{(p)}$ the thorny graph $G(p_1, p_2, \ldots, p_n)$ for short. Recall the definition of corona product, the graph $G^{(p)} \cong G \circ \overline{K_p}$ where $\overline{K_p}$ denotes the complement of complete graph K_p. Therefore, for a connected graph G of order n, we have

Example 4.1.5 ([8]) $H(G^{(p)}) = H(G) + pH_1(G) + p^2 H_2(G_1) + \dfrac{1}{4}(p + 3)np.$

Theorem 4.1.6 ([8]) *Let G_i be a connected graph of order n_i for $i = 1, 2$ and the diameter of G_2 be d. Then we have*

$$A + n_2(n_2 - 1)H_d(G_1) \leq H(G_1 \times G_2) \leq A + n_2(n_2 - 1)H_1(G_1)$$

where $A = n_1 H(G_2) + n_2 H(G_1)$. Moreover, both equalities above hold if and only if $G_2 \cong K_{n_2}$.

Recently, in [10], the exact formula of the Harary index of $G_1 \times G_2$ was given in which the Wiener polynomial of a graph is involved.

By choosing $G_1 = G$ and $G_2 = K_2$, we can easily arrive at the following corollary:

Corollary 4.1.7 ([8]) *Let G be a connected graph of order n. Then*

$$H(G \times K_2) = 2H(G) + n + 2H_1(G).$$

The lattice graph $L_{2,n}$ (see [11]) is only $P_n \times K_2$. For example, the lattice graph $L_{2,5}$ is shown in Fig. 4.1. It is well known that [12] $H(P_n) = n \sum_{i=1}^{n-1} \frac{1}{i} - n + 1$. So we have

Example 4.1.8 ([8])

$$H(L_{2,n}) = 2H(P_n) + n + 2H_1(P_n)$$

$$= 4n \sum_{i=3}^{n-1} \frac{1}{i} + n + 6.$$

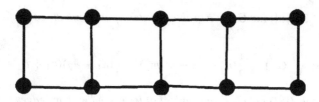

Fig. 4.1 The lattice graph $L_{2,5}$

Theorem 4.1.9 ([8]) *Let G_i be a connected graph of order n_i and with m_i edges for $i = 1, 2$. Then*

$$H(G_1[G_2]) = \frac{1}{4}n_1n_2(n_2 - 1) + \frac{1}{2}n_1m_2 + n_2^2 H(G_1).$$

The double graph of a given graph G, denoted by $G^{\odot}$, is constructed by making two copies of G (including the initial edge set of each), denoted by G_1 and G_2, and adding edges u_1v_2 and u_2v_1 for every edge uv of G. From the definition of lexicographic product we conclude that $G^{\odot} \cong G[K_2]$ for any connected graph G. Therefore, the following corollary can be easily obtained.

Corollary 4.1.10 ([8]) *Let G be a connected graph. Then*

$$H(G[K_2]) = \frac{3|G|}{2} + 4H(G).$$

For a complete graph K_n, we find that $K_n^{\odot}$ is a graph obtained by deleting a perfect match from the complete graph K_{2n}, which is the well-known cocktail party graph (see [13]). For example, the graph $K_3^{\odot}$ is shown in Fig. 4.2.

Example 4.1.11 ([8]) $H(K_n^{\odot}) = \dfrac{3n}{2} + 4H(K_n) = 2n^2 - \dfrac{n}{2}.$

Fig. 4.2 The graph $K_3^{\odot}$

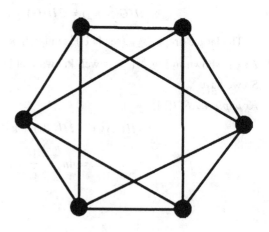

Theorem 4.1.12 ([8]) *Let G_i be a connected graph of order n_i and with m_i edges for $i = 1, 2$. Then*

$$H(G_1 \otimes G_2) = \frac{1}{4}n_1n_2(n_1n_2 - 1) + \frac{1}{2}\left(m_1n_2^2 + m_2n_1^2\right) - m_1m_2.$$

The construction of the extended double cover was introduced by Alon [14] in 1986. For a simple graph G with vertex set $V = \{v_1, v_2, \ldots, v_n\}$, the extended double cover of G, denoted by $G^\star$, is the bipartite graph with bipartition $(X; Y)$ where $X = \{x_1, x_2, \ldots, x_n\}$ and $Y = \{y_1, y_2, \ldots, y_n\}$, in which x_i and y_j are adjacent if and only if $i = j$ or v_i and v_j are adjacent in G. Note that, for a graph G, $G^\star \cong G \otimes K_2$. So the corollary below follows immediately.

Corollary 4.1.13 ([8]) *Let G be a connected graph of order n and with m edges. Then we have*

$$H(G \otimes K_2) = \frac{1}{2}n(2n - 1) + \frac{1}{2}\left(4m + n^2\right) - m.$$

For a complete graph K_n, by the definition listed above, we find that $K_n \otimes K_2$ is just $K_{n,n}$.

Example 4.1.14 ([8]) $H(K_n \otimes K_2) = 2n^2 - n$.

4.2 Application of Harary Index in Pure Graph Theory

Before stating a main result, we first introduce a classic lemma for a traceable graph as follows:

Lemma 4.2.1 ([15]) *Let G be a nontrivial graph of order $n \geq 4$ with degree sequence $(d_1, d_2, \ldots, d_n)$ where $d_1 \leq d_2 \leq \cdots \leq d_n$. Suppose that there is no integer $k < \frac{n+1}{2}$ such that $d_k \leq k - 1$ and $d_{n-k+1} \leq n - k - 1$. Then G is traceable.*

Next, we give a sufficient condition in terms of Harary index for a graph to be traceable. Moreover, Lemma 4.2.1 plays an important role in the proof of the following theorem.

Theorem 4.2.2 ([16]) *Let G be a connected graph of order $n \geq 4$ and $G \notin \{K_1 \vee (K_{n-3} \cup 2K_1), K_2 \vee (3K_1 \cup K_2), K_4 \vee 6K_1\}$. If*

$$H(G) \geq \frac{1}{2}n^2 - \frac{3}{2}n + \frac{5}{2},$$

then G is traceable.

Recently, a sufficient condition using Harary index is proved for a connected bipartite graph to be Hamiltonian.

Theorem 4.2.3 ([17]) *Let $G[X, Y]$ be a connected bipartite graph with a bipartition $[X, Y]$ of order $|X| = |Y| = n \geq 2$. If*

$$H(G) \geq \frac{3}{2}n^2 - n + \frac{1}{2},$$

then G is Hamiltonian.

Moreover, it is pointed out [17] that the extremal graphs with respect to Harary index are cycle C_n (minimal case) and complete graph K_n (maximal case), respectively, among all Hamiltonian graphs of order n.

4.3 Application of Harary Index in Mathematical Chemistry

In this section, we report some applications of Harary index to mathematical chemistry.

In mathematical chemistry, topological indices are used for modeling physicochemical, pharmacologic, toxicologic, biological, and other properties of chemical compounds (see some related chapters in [18, 19]). In chemical graph theory, a connected graph with maximum degree at most 4 belongs to a family of chemical graphs depicting hydrocarbons [20]. As usual a vertex in a chemical graph represents an atom and an edge, a given bond in a molecule. In [21], an example was given for a hydrogen-depleted tree representing 2,3-dimethylpentane, as shown in Fig. 4.3, whose tree depicting its hydrogen-depleted carbon skeleton is given in Fig. 1.5.

Moreover, as it is well known, the Harary index can also be viewed as a graph invariant based on the reciprocal distance matrix $RD(G)$ of G [21–23]. For example, we can easily obtain [12] the reciprocal distance matrix of a tree T depicting 2-methylbutane as shown in Fig. 4.4 in the following form:

$$RD(T) = \begin{pmatrix} 0 & 1 & 0.5 & 0.33 & 0.5 \\ 1 & 0 & 1 & 0.5 & 1 \\ 0.5 & 1 & 0 & 1 & 0.5 \\ 0.33 & 0.5 & 1 & 0 & 0.33 \\ 0.5 & 1 & 0.5 & 0.33 & 0 \end{pmatrix}.$$

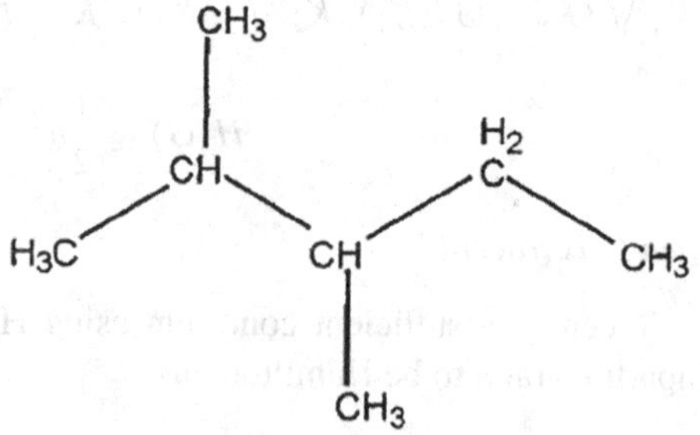

Fig. 4.3 2,3-dimethylpentane

Fig. 4.4 The tree T
depicting the carbon skeleton
of 2-methylbutane

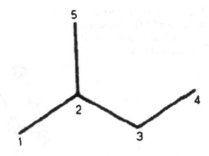

Table 4.1 The Harary
indices (H) and the Wiener
indices (W) for heptane
isomers

Heptane isomer	H	W
Heptane	11.1500	56
2-Methylhexane	11.4833	52
3-Methylhexane	11.6167	50
3-Ethylpentane	11.7500	48
2,4-Dimethylpentane	11.8333	48
2,3-Dimethylpentane	12.0000	46
2,2-Dimethylpentane	12.0833	46
3,3-Dimethylpentane	12.2500	44
2,3,3-Dimethylpentane	12.5000	42

Furthermore in [12], a table (Table 4.1) was given for comparing the Harary indices
and the Wiener indices of heptane isomers.

From this table, as pointed out in [12], Harary index appears to be a convenient
measure of the compactness of the molecule. The larger the Harary index, the larger
the compactness of the molecule. The inverse is true for the Wiener index.

In addition, the more important use of Harary index in mathematical chemistry is
in the nonempirical quantitative structure–property relationships (QSPR) and quantitative structure–activity relationships (QSAR) [12, 18, 19]. Moreover, some QSPR
studies are performed in [12] on some physical properties including boiling points
(bp), molar volumes (mv) at 20 °C, and so on, in terms of Harary index (see the next
section).

Dendrimers are nanostructures that can be precisely designed and manufactured
for a wide variety of applications, such as drug delivery, gene delivery and diag-
nostics, etc. The name dendrimer comes from the Greek word "δενδρον", which
translates to "tree". A dendrimer is generally described as a macromolecule, which
is characterized by its highly branched 3D structure that provides a high degree of
surface functionality and versatility. The first dendrimers were made by divergent
synthesis approaches by Vögtle and coworkers in 1978 [24]. Dendrimers thereafter
experienced an explosion of scientific interest because of their unique molecular
architecture.

Fig. 4.5 Molecular nanostar
dendrimer D[4] (This figure
is taken from [25] by Ashrafi,
Shabani and Diudea)

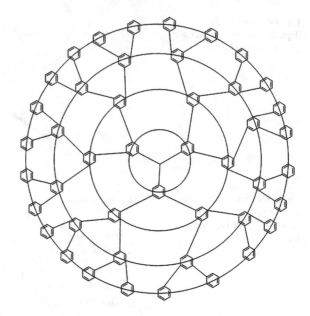

Denote by $\delta_{i,j}$ the Kronecker delta. Moreover, here we define the following notations:

$$\Delta_d = \delta_{2,d} + \delta_{3,d}, \eta(y+t) = \frac{2^{\Delta_d}}{y+t},$$

$$\mu(x,t) = 2^{x-1}\left[\frac{1}{3(x-1)+t+1} + \frac{2}{3(x-1)+t+2} + \frac{2}{3(x-1)+t+3}\right.$$
$$\left. + \frac{1}{3(x-1)+t+4}\right],$$

$$f(t,p) = \sum_{x=1}^{p}\mu(x,t) + \sum_{y=1}^{b}\eta(y+t),$$

$$p = \left[\frac{k}{4}\right] - \left(\left[\frac{m}{4}\right] - \delta_{4,d}\right) \text{ and } q = \left[\frac{m}{4}\right] - \delta_{4,d}.$$

The formula for the Harary index of molecular nanostar dendrimer D[n] (see Fig. 4.5 for the case $n = 4$) is presented in the following theorem:

Theorem 4.3.1 ([26])

$$H(D[n]) = \frac{3}{2} f\left(0, \left[\frac{k}{4}\right]\right)$$

$$+ \sum_{m=1}^{k} 3 \cdot 2^q \Big[f(|3-d|, p) + \Delta_d f(|3-d|+2, p)$$

$$+ \sum_{y=1}^{\delta_{p,-1}(k+d-2)} \frac{\eta(y)}{2^{\Delta_d}} + \delta_{2,d}(k-d) \Big]$$

$$+ \frac{3}{2} \sum_{m=1}^{k} 2^{q+\Delta_d} \Big(\sum_{z=1}^{q} f(d+3z-1, p+z) + 2f(3q+d, \left[\frac{k}{4}\right])$$

$$+ \frac{1}{3q+d} + \sum_{x=1}^{q} \frac{\mu(x, d-1)}{2^{x-1}} + \frac{10}{3} \left(2^{\left[\frac{m}{4}\right]} - 1 \right) \Big).$$

In Table 4.2, the Harary indices of D[n] are computed for $n = 1, 2, \ldots, 20$.

In [27], the Harary index of an infinite family of dendrimer nanostar NS$_2$[n] is determined. For example, NS$_2$[3] is shown in Fig. 4.6.

Table 4.2 Values of Harary index in dendrimers D[n]

n	H(D[n])	n	H(D[n])	n	H(D[n])	n	H(D[n])
1	4.5	6	163.775	11	1131.262	16	9403.148
2	19.5	7	258.443	12	1352.133	17	11508.093
3	38.9	8	291.608	13	1890.838	18	11593.067
4	56.589	9	449.681	14	3126.270	19	15747.269
5	78.168	10	788.397	15	4365.852	20	29038.085

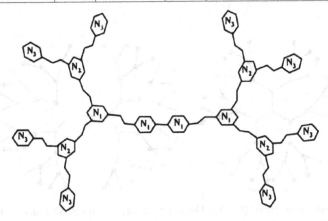

Fig. 4.6 Nanostar dendrimer NS$_2$[3] with labeled hexagons N_1, N_2, N_3

Theorem 4.3.2 ([27]) $H(\mathrm{NS}_2[n]) = H(\mathrm{NS}_2[n], x)_{x=1}$ *where* $H(\mathrm{NS}_2[n], x) = \frac{1}{l}\sum_{l=1}^{10n+9} b_l x^l$ *with b_l being the number of pairs of vertices with distance l in* $\mathrm{NS}_2[n]$.

Topology of dendrimers is basically that of a tree. Vertices in a dendrimer, except for the external endpoints, are considered as branching points. The number of edges emerging from each branching point is called *progressive degree*, that is, the edges that enlarge the number of points of a newly added generation. It equals the classical degree, k, minus one: $p = k - 1$. A regular dendrimer has all branching points with the same degree, otherwise it is irregular.

A dendrimer is called *homogeneous* if all its radial chains (i.e., the chains starting from the core and ending in an external point) have the same length [29]. The numbering of orbits (generations [30]) starts with zero for the core and ends with r, which equals the radius of the dendrimer (i.e., the number of edges from the core to the external nodes). A tree has either a monocenter or a dicenter [31] (i.e., two vertices joined by an edge). Accordingly, a dendrimer is called *monocentric* or *bicentric*. A regular monocentric dendrimer of progressive degree p and generation r is denoted by $D_{p,r}$, whereas the corresponding dicentric dendrimer, by $DD_{p,r}$. For example, two regular dendrimers $D_{2,4}$ and $DD_{2,4}$ are shown in Fig. 4.7.

Nowadays, the formulae for computing the topological indices of many molecules, especially some classes of dendrimers, are obtained for the purpose that these data of topological indices can be used for modeling the physical or chemical properties of these molecules. Many related results to this topic can be found in [28, 29, 32]. Next, we present some results on this topic. In the two theorems below, the formulae are given for the Harary indices of monocentric and bicentric dendrimers, respectively.

Theorem 4.3.3 ([33]) *Let $D_{p,r}$ be a monocentric dendrimer defined as above. Then we have*

$$H(D_{p,r}) = p \sum_{i=1}^{r} \frac{(p-1)^{i-1}}{i} + p \sum_{i=1}^{r} (p-1)^{j-1} Q_i$$

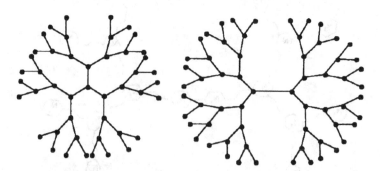

Fig. 4.7 Regular monocentric ($D_{2,4}$) and bicentric ($DD_{2,4}$) dendrimers (This figure is taken from [28] by Diudea)

where

$$Q_i = \sum_{s=1}^{i-1} \sum_{i=s}^{r+s-j} \frac{(p-1)(p-1)^{i-1}}{i+s} + \sum_{i=1}^{r-j} \frac{(p-1)^{i-1}}{i} + \sum_{i=j}^{r} \frac{(p-1)^i}{i+j}$$

$$- \sum_{s=1}^{j-1} \frac{p(p-2)(p-1)^{j+s-2}}{2s} - A_j \text{ with } A_j = \frac{p(p-1)^{2j-1}}{2j}.$$

Theorem 4.3.4 ([33]) *Let* $DD_{p,r}$ *be a dicentric dendrimer defined as above. Then we have*

$$H(DD_{p,r}) = H(D_{p,r}) + (p-1)^r \left[\sum_{i=0}^{r} J_i + \frac{1}{r+1} + (p-2) \sum_{i=1}^{r-1} \sum_{s=1}^{i} \frac{(p-1)^{s-1}}{r-i+2s} \right]$$

where

$$J_i = \frac{(p-2)(p-1)^{i-1}}{4i} + \frac{1}{r-i} + \frac{(p-1)^i}{r+i+1}.$$

Moreover, the numerical data for Harary indices of $H(D_{p,r})$ and $H(DD_{p,r})$ have been obtained [33] in Table 4.3.

In 1991, Iijima [34] first detected carbon nanotubes as multiwalled structures, which show many remarkable mechanical properties. Diudea et al. [35] were among the first chemists who considered the problem of computing the topological indices of nanostructures. As depicted in [36], for two even integers p and q, $HC_6[p,q]$ denotes an arbitrary zigzag polyhex nanotorus in terms of the circumference p and the length q (see Fig. 4.8 for the case when $p = 20$ and $q = 40$).

Note that the graph $HC_6[p,q]$ has pq vertices. Moreover, a coordinate label for the vertices of $G = HC_6[p,q]$, as shown in Fig. 4.9, was chosen [36].

In the following theorem, a formula for the Harary index of $HC_6[p,q]$ is determined completely.

Table 4.3 Harary indices of $D_{p,r}$ and $DD_{p,r}$ for some values of p and r

p	r	$H(D_{p,r})$	$H(DD_{p,r})$
3	2	22	37.86
3	3	77.6	126.48
3	4	248.23	398.43
4	3	364.3	719.54
4	4	2332	4673.5
4	6	110480	229890
6	6	25292000	66141000

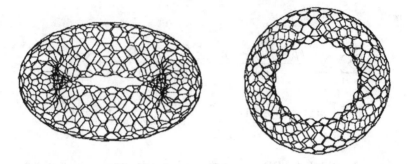

Fig. 4.8 $HC_6[20, 40]$: *Side view*; *top view* (This figure is taken from [37] by Eliasi and Taeri)

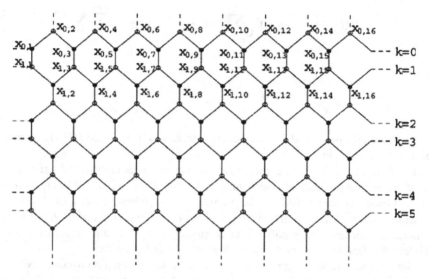

Fig. 4.9 A zigzag polyhex nanotorus lattice with $p = 16$ and $q = 6$ (This figure is taken from [37] by Eliasi and Taeri)

Theorem 4.3.5 ([36]) *Let $p = 2c$ and $q = 2d$. Suppose that $HC_6[p, q]$ is a zigzag polyhex nanotorus defined as above. Then we have*

$$H(HC_6[p, q]) = \frac{pq}{2}\left(\sum_{i=0}^{d} b_i + \sum_{j=1}^{d-1} w_j\right)$$

where

$$b_0 = 2\sum_{j=2}^{c} \frac{1}{j-1} + \frac{1}{c},$$

$$b_k = \begin{cases} 2\sum_{j=k+2}^{c} \dfrac{1}{k+j-1} + \dfrac{k+1}{2k} + \dfrac{k}{2k-1} & \text{if } k < c; \\[2ex] \dfrac{c}{2k-1} + \dfrac{c}{2k} & \text{if } c \le k < d. \end{cases}$$

for $k = 1, 2, \ldots, d$ *and*

$$w_k = \begin{cases} 2\sum_{j=k+2}^{c} \dfrac{1}{k+j-1} + \dfrac{k+1}{2k} + \dfrac{k}{2k+1} & \text{if } k < c; \\[2ex] \dfrac{c}{2k+1} + \dfrac{c}{2k} & \text{if } c \le k < d. \end{cases}$$

for $k = 1, 2, \ldots, d-1$.

Table 4.4 Wiener indices W, Harary indices H, modified Harary indices $^m H$, the number of paths of length 3 p_3, and the number carbon atoms N of 74 lower alkanes

Alkane	W	H	$^m H$	p_3	N
Ethane	1	1.0000	1.0000	0	2
Propane	4	2.5000	1.6000	0	3
Butane	10	4.3333	2.0901	1	4
2-Methylpropane	9	4.5000	2.0000	0	4
Pentane	20	6.4167	2.5254	2	5
2-Methylbutane	18	6.6667	2.4276	2	5
2,2-Dithylpropane	16	7.0000	2.2857	0	5
Hexane	35	8.7000	2.9267	3	6
2-Methylpentane	32	9.0000	2.8313	3	6
3-Methylpentane	31	9.0833	2.8159	4	6
2,2-Dimethylbutane	28	9.5000	2.6721	3	6
2,3-Dimethylbutane	29	9.3333	2.7319	4	6
Heptane	56	11.1500	3.3036	4	7
2-Methylhexane	52	11.4833	3.2123	4	7
3-Methylhexane	50	11.6165	3.1935	5	7
3-Ethylhexane	48	11.7498	3.1716	6	7
2,2-Dimethylpentane	46	12.0832	3.0545	4	7
2,3-Dimethylpentane	46	11.9998	3.0965	6	7
2,4-Dimethylpentane	48	11.8332	3.1222	4	7
3,3-Dimethylpentane	44	12.2498	3.0270	6	7
2,3,3-Trimethylbutane	42	12.4998	2.9549	6	7
Octane	84	13.7429	3.6622	5	8
2-Methylbeptane	79	14.1001	3.5746	5	8
3-Methylbeptane	76	14.2667	3.5555	6	8
4-Methylbeptane	75	14.3167	3.5517	6	8
4-Ethylhexane	72	14.4831	3.5276	7	8

In the process of writing this book, some novel results have been published on the Harary indices of nanostructures. For details see [10].

4.4 Application of Harary Index to Structure–Property Modeling

In this section we report the use of the Harary index in the quantitative structure–property modeling and also give the comparison with models based on the Wiener index and the modified Harary index. In Tables 4.4, 4.5 and 4.6, we give the Wiener indices W, Harary indices H, modified Harary indices $^m H$, and the number of paths of length 3 for the set of 74 lower alkanes. We consider seven of their physical properties: Boiling points at $20\,^\circ$C (bp), molar volumes at $20\,^\circ$C (mv), molar refractions at $20\,^\circ$C (mr), heats of vaporization at $25\,^\circ$C (hv), critical temperatures (ct), critical pressures (cp), and surface tensions at $20\,^\circ$C (st). Values of these properties are taken from Seybold et al. [38] and Needham et al. [39] given in Tables 4.7 and 4.8.

Table 4.5 The first continued part of Table 4.4

Alkane	W	H	$^m H$	p_3	N
2,2-Dimethylhexane	71	14.7665	3.4212	5	8
2,3-Dimethylhexane	70	14.7331	3.4580	7	8
2,4-Dimethylhexane	71	14.6498	3.4694	6	8
2,5-Dimethylhexane	74	14.4665	3.4887	5	8
3,3-Dimethylhexane	67	15.0331	3.3877	7	8
3,4-Dimethylhexane	68	14.8664	3.4406	8	8
3-Ethyl-2-methylpentane	67	14.9164	3.4328	8	8
3-Ethyl-3-methylpentane	64	15.2497	3.3553	9	8
2,2,3-Trimethylpentane	63	15.4164	3.3033	8	8
2,2,4-Trimethylpentane	66	15.1665	3.3368	5	8
2,3,3-Trimethylpentane	62	15.4997	3.2907	9	8
2,3,4-Trimethylpentane	65	15.1664	3.3654	8	8
2,2,3,3-Tetramethylbutane	58	15.9997	3.1657	9	8
Nonane	120	16.4606	4.0064	6	9
2-Methyloctane	114	16.8358	3.9218	6	9
3-Methyloctane	110	17.0263	3.9034	7	9
4-Methyloctane	108	17.1095	3.8986	7	9
3-Ethylheptane	104	17.3000	3.8737	8	9
4-Ethylheptane	102	17.3832	3.8667	8	9
2,2-Dimethylheptane	104	17.5502	3.7731	6	9
2,3-Dimethylheptane	102	17.5500	3.8081	8	9

Table 4.6 The second continued part of Table 4.4

Alkane	W	H	$^m H$	p_3	N
2,4-Dimethylheptane	102	17.5167	3.8156	7	9
2,5-Dimethylheptane	104	17.4167	3.8213	7	9
2,6-Dimethylheptane	108	17.2170	3.8388	6	9
3,3-Dimethylheptane	98	17.8832	3.7389	8	9
3,4-Dimethylheptane	98	17.7665	3.7865	9	9
3,5-Dimethylheptane	100	17.6332	3.8016	8	9
4,4-Dimethylheptane	96	17.9832	3.7318	8	9
3-Ethyl-2-methylhexane	96	17.8497	3.7752	9	9
4-Ethyl-2-methylhexane	98	17.7164	3.7912	8	9
3-Ethyl-3-methylhexane	92	18.2330	3.6975	10	9
3-Ethyl-4-methylhexane	94	17.9830	3.7597	10	9
2,2,3-Trimethylhexane	92	18.3497	3.6535	9	9
2,2,4-Trimethylhexane	94	18.1831	3.6737	7	9
2,2,5-Trimethylhexane	98	17.9498	3.6932	6	9
2,3,3-Trimethylhexane	90	18.4830	3.6375	10	9
2,3,4-Trimethylhexane	92	18.2330	3.6973	10	9
2,3,5-Trimethylpentane	96	17.9664	3.7269	8	9
2,4,4-Trimethylhexane	92	18.3164	3.6579	8	9
3,3,4-Trimethylhexane	88	18.6163	3.6217	11	9
3,3-Diethylpentane	88	18.4996	3.6604	12	9
2,2-Diethyl-3-ethylpentane	88	18.5830	3.6264	10	9
2,3-Diethyl-3-ethylpentane	86	18.7496	3.6023	12	9
2,4-Diethyl-3-ethylpentane	90	18.3330	3.6846	10	9
2,2,3,3-Tetramethylpentane	82	19.2497	3.4878	12	9
2,2,3,4-Tetramethylpentane	86	18.8330	3.5660	10	9
2,2,4,4-Tetramethylpentane	88	18.7498	3.5482	6	9
2,3,3,4-Tetramethylpentane	84	18.9996	3.5442	12	9

Several structure–property models were studied. Here we report statistical parameters for one-descriptor models (Table 4.9) and two-descriptor models (Table 4.10). These models were obtained by the program CROMRsel [40]. The CROMRsel computer procedure allows one to select for each property and for each model with I descriptors, the best possible model according to the highest fitted correlation coefficient. The quality of models is expressed by descriptive statistical parameters: the correlation coefficient r, the standard error of estimate s, and F-test. Models are also internally validated by computing the leave-one-out cross-validated correlation coefficient r_{cv} and standard error of estimate s_{cv}.

Results in Table 4.9 indicate that one-parameter models are not good enough, but they also show that one-parameter models based on modified Harary index $^m H$

Table 4.7   Boiling points bp/°C, molar volumes mv/cm³, molar refractions mr/cm³, heats of vaporization hv/kJ, critical temperatures ct/°C, critical pressures cp/atm, and surface tensions st/dyne cm⁻¹ of considered lower alkanes

Alkane	$\frac{bp}{°C}$	$\frac{mv}{cm^3}$	$\frac{mr}{cm^3}$	$\frac{hv}{kJ}$	$\frac{ct}{°C}$	$\frac{cp}{atm}$	$\frac{st}{dyne\ cm^{-1}}$
Ethane	−88.6				32.3	48.2	
Propane	−42.1				96.8	42.0	
Butane	−0.5				152.0	37.5	
2-Methylprotane	−11.7				135.0	36.0	
Pentane	36.1	115.2	25.27	26.4	196.6	33.3	16.00
2-Methylbutane	27.9	116.4	25.29	24.6	187.8	32.9	15.00
2,2-Dimethylprotane	9.5	122.1	25.72	21.8	160.6	31.6	
Hexane	68.7	130.7	29.91	31.6	234.7	29.9	18.42
2-Methylpentane	60.3	131.9	29.95	29.9	224.9	30.0	17.38
3-Methylpentane	63.3	129.7	29.80	30.3	231.2	30.8	18.12
2,2-Dimethylbutane	49.7	132.7	29.93	27.7	216.2	30.7	16.30
2,3-Dimethylbutane	58.0	130.2	29.81	29.1	227.1	31.0	17.37
Heptane	98.4	146.5	34.55	36.6	267.0	27.0	20.26
2-Methylhexane	90.1	147.7	34.59	34.8	257.9	27.2	19.29
3-Methylhexane	91.9	145.8	34.46	35.1	262.4	28.1	19.79
3-Ethylpentane	93.5	143.5	34.28	35.2	267.6	28.6	20.44
2,2-Dimethylpentane	79.2	148.7	34.62	32.4	247.7	28.4	18.02
2,3-Dimethylpentane	89.8	144.2	34.32	34.2	264.6	29.2	19.96
2,4-Dimethylpentane	80.5	148.9	34.62	32.9	247.1	27.4	18.15
3,3-Dimethylpentane	86.1	144.5	34.33	33.0	263.0	30.0	19.59
2,3,3-Trimethylbutane	80.9	145.2	34.37	32.0	258.3	29.8	18.76
Octane	125.7	162.6	39.19	41.5	296.2	24.64	21.76
2-Methylheptane	117.6	163.7	39.23	39.7	288.0	24.80	20.60
3-Methylheptane	118.9	161.8	39.10	39.8	292.0	25.60	21.17
4-Methylheptane	117.7	162.1	39.12	39.7	290.0	25.60	21.00
3-Ethylhexane	118.5	160.1	38.94	39.4	292.0	25.74	21.51
2,2-Dimethylhexane	106.8	164.3	39.25	37.3	279.0	25.60	19.60
2,3-Dimethylhexane	115.6	160.4	38.98	38.8	293.0	26.60	20.99
2,4-Dimethylhexane	109.4	163.1	39.13	37.8	282.0	25.80	20.05
2,5-Dimethylhexane	109.1	164.7	39.26	37.9	279.0	25.00	19.73
3,3-Dimethylhexane	112.0	160.9	39.01	37.9	290.8	27.20	20.63
3,4-Dimethylhexane	117.7	158.8	38.85	39.0	298.0	27.40	21.62
3-Ethyl-2-methylpentane	115.7	158.8	38.84	38.5	295.0	27.40	21.52
3-Ethyl-3-methylpentane	118.3	157.0	38.72	38.0	305.0	28.90	21.99

produce the best statistical parameters in five cases (bp, mv, hv, ct, cp) and Harary index H in two cases (mr, st).

In Table 4.10, we report the best two-descriptor models, obtained by using the CROMRsel procedure and parameters given in Tables 4.4, 4.5, and 4.6.

Table 4.8 The continued part of Table 4.7

Alkane	bp $^\circ$C	mv cm^3	mr cm^3	hv kJ	ct $^\circ$C	cp atm	st dyne cm^{-1}
2,2,3-Trimethylpentane	109.8	159.5	38.92	36.9	294.0	28.20	20.67
2,2,4-Trimethylpentane	99.2	165.1	39.26	36.1	271.2	25.50	18.77
2,3,3-Trimethylpentane	114.8	157.3	38.76	37.2	303.0	29.00	21.56
2,3,4-Trimethylpentane	113.5	158.9	38.87	37.6	295.0	27.60	21.14
2,2,3,3-Tetramethylbutane	106.5				270.8	24.50	
Nonane	150.8	178.7	43.84	46.4	322.0	22.74	22.92
2-Methyloctane	143.3	179.8	43.88	44.7	315.0	23.60	21.88
3-Methyloctane	144.2	178.0	43.73	44.8	318.0	23.70	22.34
4-Methyloctane	142.5	178.2	43.77	44.8	318.3	23.06	22.34
3-Ethylheptane	143.0	176.4	43.64	44.8	318.0	23.98	22.81
4-Ethylheptane	141.2	175.7	43.49	44.8	318.3	23.98	22.81
2,2-Dimethylheptane	132.7	180.5	43.91	42.3	302.0	22.80	20.80
2,3-Dimethylheptane	140.5	176.7	43.63	43.8	315.0	23.79	22.34
2,4-Dimethylheptane	133.5	179.1	43.74	42.9	306.0	22.70	21.30
2,5-Dimethylheptane	136.0	179.4	43.85	42.9	307.8	22.70	21.30
2,6-Dimethylheptane	135.2	180.9	43.93	42.8	306.0	23.70	20.83
3,3-Dimethylheptane	137.3	176.9	43.69	42.7	314.0	24.19	22.01
3,4-Dimethylheptane	140.6	175.3	43.55	43.8	322.7	24.77	22.80
3,5-Dimethylheptane	136.0	177.4	43.64	43.0	312.3	23.59	21.77
4,4-Dimethylheptane	135.2	176.9	43.60	42.7	317.8	24.18	22.01
3-Ethyl-2-methylhexane	138.0	175.4	43.66	43.8	322.7	24.77	22.80
4-Ethyl-2-methylhexane	133.8	177.4	43.65	43.0	330.3	25.56	21.77
3-Ethyl-3-methylhexane	140.6	173.1	43.27	43.0	327.2	25.66	23.22
3-Ethyl-4-methylhexane	140.4.6	172.8	43.37	44.0	312.3	23.59	23.27
2,2,3-Trimethylhexane	133.6	175.9	43.62	41.9	318.1	25.07	21.86
2,2,4-Trimethylhexane	126.5	179.2	43.76	40.6	301.0	23.39	20.51
2,2,5-Trimethylhexane	124.1	181.3	43.94	40.2	296.6	22.41	20.04
2,3,3-Trimethylhexane	137.7	173.8	43.43	42.2	326.1	25.56	22.41
2,3,4-Trimethylhexane	139.0	173.5	43.39	42.9	324.2	25.46	22.80
2,3,5-Trimethylhexane	131.3	177.7	43.65	41.4	309.4	23.49	21.27
2,4,4-Trimethylhexane	130.6	177.2	43.66	40.8	309.1	23.79	21.17
3,3,4-Trimethylhexane	140.5	172.1	43.34	42.3	330.6	26.45	23.27
3,3-Diethylpentane	146.2	170.2	43.11	43.4	342.8	26.94	23.75
2,2-Dimethyl-3-ethylpentane	133.8	174.5	43.46	42.0	338.6	25.96	22.38
2,3-Dimethyl-3-ethylpentane	142.0	170.1	42.95	42.6	322.6	26.94	23.87
2,4-Dimethyl-3-ethylpentane	136.7	173.8	43.40	42.9	324.2	25.46	22.80
2,2,3,3-Tetramethylpentane	140.3	169.5	43.21	41.0	334.5	27.04	23.38
2,2,3,4-Tetramethylpentane	133.0	173.6	43.44	41.0	319.6	25.66	21.98
2,2,4,4-Tetramethylpentane	122.3	178.3	43.87	38.1	301.6	24.58	20.37
2,3,3,4-Tetramethylpentane	141.6	169.9	43.20	41.8	334.5	26.85	23.31

Table 4.9 Statistical parameters for the best one-descriptor models

$bp = -145.30(\pm 2.75) + 68.03(\pm 1.06)^m H + 3.51(\pm 0.20)p_3$ $N = 74 \quad r = 0.9974 \quad r_{cv} = 0.9968 \quad s = 3.67 \quad s_{cv} = 3.31$
$mv = 30.39(\pm 0.79) + 17.79(\pm 0.13)N - 1.71(\pm 0.06)p_3$ $N = 69 \quad r = 0.9990 \quad r_{cv} = 0.9989 \quad s = 0.75 \quad s_{cv} = 0.80$
$mr = 1.67(\pm 0.06) + 4.78(\pm 0.01)N - 0.127(\pm 0.004)p_3$ $N = 69 \quad r = 0.9999 \quad r_{cv} = 0.9999 \quad s = 0.054 \quad s_{cv} = 0.058$
$hv = -6.14(\pm 0.64) + 12.27(\pm 0.22)^m H - 0.36(\pm 0.03)p_3$ $N = 69 \quad r = 0.9953 \quad r_{cv} = 0.9885 \quad s = 0.51 \quad s_{cv} = 0.54$
$ct = -8.17(\pm 6.27) + 73.05(\pm 2.43)^m H - 6.62(\pm 0.46)p_3$ $N = 74 \quad r = 0.9911 \quad r_{cv} = 0.9989 \quad s = 7.56 \quad s_{cv} = 8.62$
$cp = 51.43(\pm 0.74) - 3.77(\pm 0.14)N - 0.79(\pm 0.08)p_3$ $N = 74 \quad r = 0.9726 \quad r_{cv} = 0.9656 \quad s = 1.00 \quad s_{cv} = 1.12$
$st = 8.28(\pm 0.64) + 2.73(\pm 0.22)^m H - 0.45(\pm 0.03)p_3$ $N = 68 \quad r = 0.9667 \quad r_{cv} = 0.9632 \quad s = 0.49 \quad s_{cv} = 0.51$

Table 4.10 The best two-descriptor models obtained by the program CROMRsel

Descriptor	r	r_{cv}	s	s_{cv}	F-test
bp-74					
W	0.9170	0.9057	18.341	19.495	381
H	0.9564	0.9502	13.426	14.333	773
$^m H$	0.9863	0.9858	7.575	7.737	2582
mv-69					
W	0.9720	0.9693	4.027	4.213	1145
H	0.9586	0.9562	4.977	5.012	760
$^m H$	0.9638	0.9608	4.567	4.752	876
mr-69					
W	0.9616	0.9581	1.430	1.494	823
H	0.9817	0.9805	0.993	1.024	1778
$^m H$	0.9612	0.9588	1.438	1.481	813
hv-69					
W	0.9642	0.9609	1.412	1.476	887
H	0.9097	0.9033	2.212	2.286	322
$^m H$	0.9870	0.9864	0.856	0.874	2531
ct-74					
W	0.8832	0.8682	26.666	28.221	255
H	0.9560	0.9489	16.672	17.945	765
$^m H$	0.9653	0.9635	14.838	15.231	985
cp-74					
W	0.8729	0.8532	2.100	2.246	231
H	0.8730	0.8524	2.099	2.251	231
$^m H$	0.9610	0.9864	1.190	1.281	870
st-68					
W	0.8109	0.7965	1.120	1.158	127
H	0.8687	0.8594	0.948	0.979	203
$^m H$	0.8534	0.8451	0.998	1.023	177

The results in Table 4.10 show that the most often used descriptor in all two-parameter models is the number of paths which is involved in all models. The modified Harary index is used in four models (bp, hv, ct, st) and the number of carbon atoms N is three models (mv, mr, cp). Therefore, it seems that N is a valuable descriptor for the structure–property modeling of lower alkanes. However, N cannot distinguish isomers, producing the so-called comb-type correlations. Thus, the models based on the combinations of two descriptors are prefered if one molecular descriptor is the Harary index or the modified Harary index. For example, the prefered model for molar volume should be one based on the combination of the Harary index and N with the following statistical parameters: $r = 0.9874$, $r_{cv} = 0.9864$, $s = 2.713$, $s_{cv} = 2.816$.

References

1. Dirac GA (1952) Some theorems on abstract graphs. Proc Lond Math Soc 2:69–81
2. Ore O (1960) Note on Hamiltonian circuits. Am Math Mon 67:55
3. Fan G (1984) New sufficient conditions for cycles in graphs. J Combin Theory Ser B 37:221–227
4. Gould RJ (1991) Updating the Hamiltonian problem—a survey. J Graph Theory 15:121–157
5. Xu K, Das KC (2011) On Harary index of graphs. Discret Appl Math 159:1631–1640
6. Klavžar S (1996) Coloring graph products-a survey. Discret Math 155:135–145
7. Hammack R, Imrich W, Klavžar S (2011) Handbook of product graphs, 2nd edn. CRC Press, Boca Raton
8. Das KC, Xu K, Cangul IN, Cevik AS, Graovac A (2013) On the Harary index of graph operations. J Inequal Appl 2013:1–16
9. Gutman I (1998) Distance in thorny graph. Publ Inst Math (Beograd) 63:31–36
10. Azari M, Iranmanesh A (2014) Harary index of some nano-structures. MATCH Commun Math Comput Chem 71:373–382
11. Brouwer AE, Cohen AM, Neumaier A (1989) Distance-regular graphs. Springer, Berlin
12. Plavšić D, Nikolić S, Trinajstić N, Mihalić Z (1993) On the Harary index for the characterization of chemical graphs. J Math Chem 12:235–250
13. Biggs NL (1993) Distance-transitive graphs. Algebraic graph theory, 2nd edn. Cambridge University Press, Cambridge, pp 155–163
14. Alon N (1986) Eigenvalues and expanders. Combinatorica 6:83–96
15. Bondy JA, Murty USR (1976) Graph theory with applications. Macmillan Press, New York
16. Hua H, Wang M (2013) On Harary index and traceable graphs. MATCH Commun Math Comput Chem 70:297–300
17. Zeng T (2013) Harary index and Hamiltonian property of graphs. MATCH Commun Math Comput Chem 70:645–649
18. Todeschini R, Consonni V (2000) Handbook of molecular descriptors. Wiley-VCH, Weinheim
19. Todeschini R, Consonni V (2009) Molecular descriptors for chemoinformatics, vol I, vol II. Wiley-VCH, Weinheim, pp 934–938
20. Trinajstić N (1992) Chemical graph theory. CRC Press, Boca Raton
21. Lučić B, Miličević A, Nikolić S, Trinajstić N (2002) Harary index-twelve years later. Croat Chem Acta 75:847–868
22. Janežič D, Miličević A, Nikolić S, Trinajstić N (2007) Graph theoretical matrices in chemistry. University of Kragujevac, Kragujevac
23. Lučić B, Sović I, Plavšić D, Trinajstić N (2012) Harary matrices: definitions, properties and applications. In: Gutman I, Furtula B (eds) Distance in molecular graphs-applications. University of Kragujevac, Kragujevac, pp 3–26

24. Buhleier E, Wehner W, Vögtle F (1978) Cascade and nonskid-chain-like synthesis of molecular cavity topologies. Synthesis 1978:155–158
25. Ashrafi AR, Shabani H, Diudea MV (2013) Balaban index of dendrimers. MATCH Commun Math Comput Chem 69:151–158
26. Yavari N, Shabani H, Fazlollahi HR, Diudea MV (2014) Computing the Harary index of a class of nanostar dendrimers. Stud Univ Babes-Bolyai Chem 59:201–205
27. Alikhani S, Iranmanesh MA, Taheri H (2014) Harary index of dendrimer nanostar $NS_2[n]$. MATCH Commun Math Comput Chem 71:383–394
28. Diudea MV (1995) Molecular topology 21. Wiener indices of dendrimers. MATCH Commun Math Comput Chem 32:71–83
29. Diudea MV, Kiss AA, Estrada E, Guavara N (2000) Connectivity-, Wiener- and Harary-type indices of dendrimers. Croat Chem Acta 73:367–381
30. Tomalia DA, Naylor AM, Goddard WA (1990) Starburst dendrimers: molecular-level control of size, shape, surface chemistry, topology, and flexibility from atoms to macroscopic matter. Angew Chem Int Ed Engl 29:138–175
31. Harary F (1969) Graph theory. Addison-Wesley, Reading
32. Xu K (2011) Computing the Hosoya index and the Wiener index of an infinite class of dendrimers. Dig J Nanomater Biostructures 6:265–270
33. Heydari A (2010) Harary index of regular dendrimers. Optoelectron Adv Mater-Rapid Commun 4:2206–2208
34. Iijima S (1991) Helical microtubules of graphitic carbon. Nature 354:56–58
35. Diudea MV, Silaghi-Dumitrescu I, Parv B (2001) Toranes versus torenes. MATCH Commun Math Comput Chem 44:117–133
36. Eliasi M (2009) Harary index of zigzag polyhex nanotorus. Dig J Nanomater Biostructures 4:755–760
37. Eliasi M, Taeri B (2008) Hosoya polynomial of zigzag polyhex nanotorus. J Serb Chem Soc 73:311–319
38. Seybold PG, May M, Bagal UA (1987) Molecular structure-property relationships. J Chem Educ 64:575–581
39. Needham DE, Wei IC, Seybold PG (1988) Molecular modeling of the physical properties of alkanes. J Am Chem Soc 110:4186–4194
40. Lučić B, Trinajstić N (1999) Multivariate regression outperforms several robust architectures of neural networks in QSAR modeling. J Chem Inf Comput Sci 39:121–132

Chapter 5
The Variants of Harary Index

Nowadays, several variants of Harary index are introduced from the theoretical or applied viewpoint [1–4]. As an instance, to determine the upper or lower bounds on Harary index of graphs, we define a new variant of Harary index as follows [5]

$$H_t(G) = \sum_{u,v \in V(G)} \frac{1}{d_G(u,v) + t}$$

where the summation goes over all unordered pairs of vertices in graph G. Moreover, as another variant of Harary index, the normalized Harary index of graph G of order n, denoted by NH(G), is defined [6] as

$$\text{NH}(G) = \frac{H(G) - H(P_n)}{\frac{n(n+5)}{4} - H(P_n)}.$$

Other related results to normalize Harary index can be found in [7]. For some details of other variants of Harary index of a graph, please see [2].

Recently, two variants of Harary index were introduced [8], which are additively weighted Harary index and multiplicatively weighted Harary index. For a graph G, the additively weighted Harary index is defined [8] as follows:

$$H_A(G) = \sum_{\{u,v\} \subseteq V(G)} \frac{d_G(u) + d_G(v)}{d_G(u,v)}.$$

Moreover, the multiplicatively weighted Harary index of graph G is defined [8] in the following form:

$$H_M(G) = \sum_{\{u,v\} \subseteq V(G)} \frac{d_G(u)d_G(v)}{d_G(u,v)}.$$

In the meantime, Hua and Zhang [9] gave another name—reciprocal degree distance (denoted by RDD), to the additively weighted Harary index. Actually, this

© The Author(s) 2015

K. Xu et al., *The Harary Index of a Graph*,
SpringerBriefs in Mathematical Methods, DOI 10.1007/978-3-662-45843-3_5

name is derived from another well-known topological index—degree distance (see [10, 11]) in chemical graph theory as its degree version, whereas the degree distance of graph G is defined as:

$$\mathrm{DD}(G) = \sum_{\{u,v\} \subseteq V(G)} (d_G(u) + d_G(v)) d_G(u, v).$$

In this chapter we report some interesting results on these two variants of Harary index, i.e., the additively and multiplicatively weighted Harary indices, by mainly focusing on extremal graphs with respect to them and some properties of these two variants. For the sake of consistency, we always denote by $H_A(G)$ and $H_M(G)$ the additively weighted Harary index and the multiplicatively weighted Harary index of graph G, respectively.

5.1 Extremal Graphs with Respect to H_A and H_M

In this section, we present some extremal results with respect to the additively and multiplicatively weighted Harary indices for the graphs from various sets of graphs.

In the following, we first deal with the extremal results for the general graphs and then consider the ones for special graphs.

Theorem 5.1.1 ([9]) *Among all connected graphs of order n, the complete graph K_n has the maximum additively weighted Harary index, and the path P_n has the minimum additively weighted Harary index.*

In the next theorem, we determine the extremal graphs with maximal additively weighted Harary index among all cacti with k cycles. Recall that $C^0(n, k)$ is a special cactus as defined in Sect. 2.1.

Theorem 5.1.2 ([9]) *Let G be a cactus of order n and with k cycles where $0 \leq k \leq \dfrac{n-1}{2}$. Then, we have*

$$H_A(G) \leq \frac{3n^2 + (2k - 5)n + 4k + 2}{2}$$

with equality holding if and only if $G \cong C^0(n, k)$.

Note that KC_n^k is a graph obtained by attaching k pendant vertices to one vertex of complete graph K_{n-k} as defined in Sect. 2.1.

Theorem 5.1.3 ([9]) *Let G be a connected graph of order n and with k pendant vertices. Then, we have*

$$H_A(G) \leq \frac{3n^3 - 8pn^2 + (7p^2 + 17p + 3)n - 2p^3 - 7p^2 - 7p - 2}{4}$$

with equality holding if and only if $G \cong KC_n^k$.

Theorem 5.1.4 ([9]) *Let G be a connected graph of order n and with independence number α. Then, we have*

$$H_A(G) \leq n^3 - (\alpha + 1)n^2 - \left(\frac{3}{2}\alpha^2 - \frac{3}{2}\alpha - 1\right)n + \frac{1}{2}\alpha^3 + \frac{1}{2}\alpha - \alpha$$

with equality holding if and only if $G \cong \overline{K_\alpha} \bigvee K_{n-\alpha}$.

In the theorem below, the extremal graph with maximal additively weighted Harary index is characterized among all graphs with a given chromatic number.

Theorem 5.1.5 ([9]) *Let G be a connected graph of order n and with chromatic number k such that $n = pk + p$ where $0 \leq q \leq p - 1$. Then, we have*

$$H_A(G) \leq n^3 - (3q + 2)n^2 + \left(\frac{3}{2}q^2 k + \frac{3}{2}qk + \frac{3}{2}q^2 + \frac{5}{2}q + 1\right)n - q(q + 1)^2 k$$

with equality holding if and only if G is isomorphic to Turán graph $T_n(k)$.

Theorem 5.1.6 ([9]) *Let G be a connected graph of order n and with vertex-(or edge-) connectivity k. Then, we have*

$$H_A(G) \leq n^3 - \frac{9}{2}n^2 + \left(2k + \frac{13}{2}\right)n + \frac{1}{2}k^2 - \frac{5}{2}k - 3$$

with equality holding if and only if $G \cong K_k \bigvee (K_1 \cup K_{n-k-1})$.

Note that $f(k) = n^3 - \frac{9}{2}n^2 + \left(2k + \frac{13}{2}\right)n + \frac{1}{2}k^2 - \frac{5}{2}k - 3$ is an strictly increasing function. Therefore, the corollary below follows immediately.

Corollary 5.1.7 ([9]) *Let G be a connected graph of order n and with vertex-(or edge-) connectivity at most k. Then, we have*

$$H_A(G) \leq n^3 - \frac{9}{2}n^2 + \left(2k + \frac{13}{2}\right)n + \frac{1}{2}k^2 - \frac{5}{2}k - 3$$

with equality holding if and only if $G \cong K_k \bigvee (K_1 \cup K_{n-k-1})$.

Theorem 5.1.8 ([12]) *Let G be a connected graph with $n \geq 4$ vertices and matching number β, where $2 \leq \beta \leq \left\lfloor \frac{n}{2} \right\rfloor$. Set $a = \dfrac{23 - 4n + \sqrt{37n^2 - 121n + 109}}{21}$*

(1) *If $\beta = \left\lfloor \frac{n}{2} \right\rfloor$, then $H_A(G) \leq n(n-1)^2$ with equality holding if and only if $G \cong K_n$;*

(2) *If $a < \beta \leq \left\lfloor \frac{n}{2} \right\rfloor - 1$, then $H_A(G) \leq 4\beta^3 + (2n - 12)\beta^2 + (11 - 3n)\beta + \dfrac{n^2 - n - 4}{2}$ with equality holding if and only if $G \cong K_1 \bigvee (K_{2\beta-1} \cup \overline{K_{n-2\beta}})$;*

(3) *If* $2 \leq \beta < a$, *then* $H_A(G) \leq \dfrac{1}{2}\beta^3 - \dfrac{1}{2}\beta^2 + \dfrac{n^2 - 3n + 2}{2}\beta$ *with equality holding if and only if* $G \cong K_\beta \bigvee \overline{K_{n-\beta}}$;

(4) *If* $\beta = a$, *then* $H_A(G) \leq 4a^3 + (2n - 12)a^2 + (11 - 3n)a + \dfrac{n^2 - n - 4}{2} = \dfrac{1}{2}a^3 - \dfrac{1}{2}a^2 + \dfrac{n^2 - 3n + 2}{2}a$ *with equality holding if and only if* $G \cong K_\beta \bigvee \overline{K_{n-\beta}}$ *or* $G \cong K_1 \bigvee (K_{2\beta-1} \cup \overline{K_{n-2\beta}})$.

Now, we present the following theorem in which the extremal trees are characterized completely with respect to additively weighted Harary index. Moreover, the maximal case is also determined in [9].

Theorem 5.1.9 ([8, 9]) *Let T be a tree of order n. Then, we have*

$$H_A(P_n) \leq H_A(T) \leq H_A(S_n).$$

Recall that, for $m \geq n \geq 3$, (n, m)-graph is a connected graph of order n and with m edges. Let $\mathcal{G}_{n,m}$ denote the set of (n, m)-graphs. For $n \geq 3$ and $3 \leq m \leq 2n-4$ let $G_{n,m} \in \mathcal{G}_{n,m}$ be the graph shown in Fig. 5.1. Moreover, a graph $G'_{n,n+2}$ is shown in Fig. 5.2. In the next step, we give a more general extremal result on the additively weighted Harary index, in which the (n, m)-graphs maximizing the value of H_A are completely determined.

Theorem 5.1.10 ([13]) *If $G \in \mathcal{G}_{n,m}$, $4 \leq n \leq m \leq 2n - 4$, then*

$$H_A(G) \leq \frac{m(m + 5) + 2(n - 1)(n - 3)}{2},$$

Fig. 5.1 The graphs $G_{n,m}$

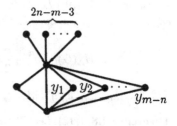

Fig. 5.2 The graphs $G'_{n,n+2}$

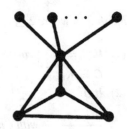

where the equality holds if and only if $G \cong G_{n,m}$ for $n \leq m \leq n+1$ or $n+3 \leq m \leq 2n-4$; and $G \cong G_{n,m}$ or $G'_{n,n+2}$ for $m = n+2$.

Assume that $n-1 = kq + r$ with $0 \leq r < k$, that is, $q = \left\lfloor \dfrac{n}{k} \right\rfloor$. Clearly, we have

$$n-1 = k\left\lfloor \frac{n}{k} \right\rfloor + r = (k-r)\left\lfloor \frac{n}{k} \right\rfloor + r\left\lceil \frac{n}{k} \right\rceil.$$

Theorem 5.1.11 ([14]) *Among all trees with n vertices and k pendant vertices, where $3 \leq k \leq n-2$, the tree $T_n\left(\left\lfloor \dfrac{n}{k} \right\rfloor^r, \left\lfloor \dfrac{n}{k} \right\rfloor^{k-r}\right)$ has the maximal additively weighted Harary index.*

Theorem 5.1.12 ([14]) *Let T be a tree with n vertices and matching number $2 \leq \beta \leq \left\lfloor \dfrac{n}{2} \right\rfloor$. Then*

$$H_A(T) \leq \frac{3}{2}n^2 + \frac{5}{12}\beta^2 - \frac{4}{3}\beta n - \frac{7}{6}n + \frac{33}{12}\beta - \frac{13}{6}$$

with equality holding if and only if $T \cong T_n(2^{\beta-1}, 1^{n-2\beta+1})$.

Theorem 5.1.13 ([14]) *Let T be a tree with n vertices and domination number $2 \leq \gamma \leq \left\lfloor \dfrac{n}{2} \right\rfloor$. Then*

$$H_A(T) \leq \frac{3}{2}n^2 + \frac{5}{12}\gamma^2 - \frac{4}{3}\gamma n - \frac{7}{6}n + \frac{33}{12}\gamma - \frac{13}{6}$$

with equality holding if and only if $T \cong T_n(2^{\gamma-1}, 1^{n-2\gamma+1})$.

For a connected graph G of order n, if its vertex set can be partitioned into two subsets V_1 and V_2, such that each edge joins a vertex in V_1 with a vertex in V_2 with $|V_1| = p \leq |V_2| = q$ and $p + q = n$. Then we say that G has a (p, q)-bipartition. In the following theorem, the extremal tree with a given bipartition has been characterized completely.

Theorem 5.1.14 ([14]) *Let T be a tree of order n and with a (p, q)-bipartition where $p > 2$. Then we have*

$$H_A(T) \leq \frac{3}{2}(p^2 + q^2) + \frac{5}{3}pq - \frac{7}{6}n - \frac{1}{3}$$

with equality holding if and only if $T \cong DS_{p-1, q-1}$.

The extremal unicyclic graphs with respect to additively weighted Harary index are determined in the following theorem.

Theorem 5.1.15 ([9, 15]) *Let G be a uncyclic graph of order $n \geq 5$. Then we have*

$$4\sum_{i=1}^{n-2} H_{n-i-1} + H_{n-3} + 3H_{n-2} + \frac{6n-13}{n-2} \leq H_A(G) \leq \frac{3}{2}(n^2 - n + 2)$$

where $H_n = \sum_{i=1}^{n} \frac{1}{i}$. Moreover, the left equality holds if and only if $G \cong C_3$ $((n-3)^1)$, and the right equality holds if and only if $G \cong C_3(1^{n-3})$.

In the several theorems below we present the extremal graphs with respect to multiplicatively weighted Harary index among all trees and uncyclic graphs, respectively.

Theorem 5.1.16 ([16]) *Let T be a tree of order n. Then we have*

$$H_M(P_n) \leq H_M(T) \leq H_M(S_n)$$

with left equality holding if and only if $T \cong P_n$ and with right equality holding if and only if $T \cong S_n$.

Theorem 5.1.17 ([17]) *Let T be a tree with n vertices and k pendant vertices, where $2 \leq k \leq n-2$. Then,*

$$H_M(T) \leq H_M\left(T_n\left(\left\lceil\frac{n}{k}\right\rceil^r, \left\lfloor\frac{n}{k}\right\rfloor^{k-r}\right)\right)$$

with equality holding if and only if $T \cong T_n\left(\left\lceil\frac{n}{k}\right\rceil^r, \left\lfloor\frac{n}{k}\right\rfloor^{k-r}\right)$.

Theorem 5.1.18 ([17]) *Let T be a tree with n vertices and matching number $\beta \leq \frac{n}{2}$. Then*

$$H_M(T) \leq \frac{5}{4}n^2 - \frac{3}{8}\beta^2 - \frac{1}{6}n\beta - \frac{31}{12}n + \frac{13}{12}\beta + \frac{1}{4}$$

with equality holding if and only if $T \cong T_n(2^{\beta-1}, 1^{n-2\beta+1})$.

Theorem 5.1.19 ([17]) *Let T be a tree with n vertices and domination number γ. Then*

$$H_M(T) \leq \frac{5}{4}n^2 - \frac{3}{8}\gamma^2 - \frac{1}{6}n\gamma - \frac{31}{12}n + \frac{13}{12}\gamma + \frac{1}{4}$$

with equality holding if and only if $T \cong T_n(2^{\gamma-1}, 1^{n-2\gamma+1})$.

Theorem 5.1.20 ([17]) *Among all trees of order n and with diameter d, the tree $T_n\left(\left\lceil\frac{d}{2}\right\rceil, \left\lfloor\frac{d}{2}\right\rfloor, 1^{n-d-1}\right)$ maximizes the multiplicatively weighted Harary index.*

Theorem 5.1.21 ([17]) *Let T be a tree of order n and with a (p,q)-bipartition where $p > 2$. Then we have*

$$H_A(T) \leq \frac{5}{4}(p^2 + q^2) + \frac{7}{3}pq - \frac{31}{12}n + \frac{4}{3}$$

with equality holding if and only if $T \cong \mathrm{DS}_{p-1,q-1}$.

Theorem 5.1.22 ([16]) *Let G be a uncyclic graph of order $n \geq 3$. Then we have*

$$H_M(G) \leq \frac{n(5n + 1)}{4}.$$

Moreover, the equality holds if and only if $G \cong C_3(1^{n-3})$.

Below we present two more general results on the extremal graphs with respect to multiplicatively weighted Harary index. Let K_k^{n-k} be the graph obtained from the complete graph K_k by attaching $n - k$ pendent vertices to one vertex of K_k. Clearly, K_2^{n-2} is the star S_n of order n and K_3^{n-3} is the graph obtained by inserting a new edge between two pendent vertices of S_n. Moreover, from Fig. 5.2 we find that $K_4^{n-4} \cong G'_{n,n+2}$.

Theorem 5.1.23 ([13]) *If $G \in \mathcal{G}_{n,n+1}$, $n \geq 4$, then*

$$H_M(G) \leq \frac{5n^2 + 13n + 8}{4}$$

with equality holding if and only if $G \cong G_{n,n+1}$.

Theorem 5.1.24 ([13]) *If G is an $(n, n + \binom{k}{2} - k)$-graph, $n \geq 5$, $2 \leq k \leq n - 1$, then*

$$H_M(G) \leq \frac{(2n + k^2 - 3k)(n + k^2 - 3k + 1)}{2} + \frac{(n - k)(n + k - 3)}{4}$$

with equality holding if and only if $G \cong K_k^{n-k}$.

By choosing $k = 3$ in Theorem 5.1.24, we can directly obtain the result in Theorem 5.2.16.

Recall that the nontrivial quasi-tree graphs and k-generalized are defined in Sect. 2.1. In the last part of this section, we present several extremal results with respect to H_A and H_M, respectively, among these two sets of graphs. Note that these extremal graphs here have a slightly different form from those in [18]. The nth harmonic number is $H_n = \sum_{k=1}^{n} \frac{1}{k}$.

Theorem 5.1.25 ([18]) *Let G be a nontrivial quasi-tree graph of order $n \geq 4$. Then we have*

(1) $H_A(G) \leq 3n^2 - 5n$ *with equality holding if and only if* $G \cong K_2 \bigvee \overline{K_{n-2}}$;
(2) $H_M(G) \leq 6n^2 - 19n + 15$ *with equality holding if and only if* $G \cong K_2 \bigvee \overline{K_{n-2}}$.

Theorem 5.1.26 ([18]) *Let G be a nontrivial quasi-tree graph of order $n \geq 7$. Then we have*

(1) $H_A(G) \geq 4 \sum_{i=1}^{n-2} H_i + 4H_{n-2} + 6 - \dfrac{2}{n-2}$ *with equality holding if and only if* $G \cong C_3((n-3)^1)$;

(2) $H_M(G) \geq 4 \sum_{i=1}^{n-2} H_i + 4H_{n-3} - \dfrac{1}{n-3} + 8$ *with equality holding if and only if* $G \cong C_3((n-3)^1)$.

Theorem 5.1.27 ([18]) *For any graph $G \in QT^{(k)}(n)$ with $k \geq 2$ and $n \geq 6$, we have*

(1)
$$H_A(G) \leq \frac{(k+1)(3n^2 - 5n - k^2 - k + 2)}{2}$$

with equality holding if and only if $G \cong K_{k+1} \bigvee \overline{K_{n-k-1}}$;
(2)

$$H_M(G) \leq \frac{(k+1)^2(7n^2 - 6nk - 15n + k^2 + 7k + 8)}{4} - \frac{(k+1)(n-1)^2}{2}$$

with equality holding if and only if $G \cong K_{k+1} \bigvee \overline{K_{n-k-1}}$.

Theorem 5.1.28 ([18]) *Let G be a 2-generalized quasi-tree graph of order $n \geq 4$. Then we have*

(1)
$$H_A(G) \geq 4 \sum_{i=1}^{n-3} H_i + 9H_{n-3} + H_{n-4} + \frac{3}{n-3} + \frac{2}{n-4} + 12$$

with equality holding if and only if $G \cong C_{3,3}^{n-5}$;
(2)
$$H_M(G) \geq 4 \sum_{i=1}^{n-3} H_i + 12H_{n-3} + \frac{4}{n-4} + \frac{1}{n-5} + 16$$

with equality holding if and only if $G \cong C_{3,3}^{n-5}$.

But, unfortunately, it seems a bit difficult to determine the extremal k-generalized quasi-tree graphs for $k \geq 3$ with minimal H_A or minimal H_M, respectively.

5.2 Some Properties of Additively Weighted Harary Index

In this section, we report some properties of additively weighted Harary index of graphs, mainly including the relationship between itself and other graph invariants and the formulae for computing the additively weighted Harary indices of some graph products.

Before stating the main results, we first introduce some necessary definitions. For a graph G, the first and second Zagreb coindices of graph G are defined [19, 20] in the following:

$$\overline{M}_1 = \overline{M}_1(G) = \sum_{u \neq v, uv \notin E(G)} (d_G(u) + d_G(v)),$$

$$\overline{M}_2 = \overline{M}_2(G) = \sum_{u \neq v, uv \notin E(G)} d_G(u)d_G(v).$$

In the next step, we present some relations of $H_A(G)$ with other topological indices of graphs.

Theorem 5.2.1 ([9]) *Let G be a connected graph of order n. Then we have*

$$H_A(G) \leq M_1(G) + \overline{M}_1(G)$$

with equality holding if and only if $G \cong K_n$.

Theorem 5.2.2 ([9]) *Let G be a connected graph of order n. Then we have*

$$H_A(G) \geq \frac{\left(M_1(G) + \overline{M}_1(G)\right)^2}{DD(G)}$$

with equality holding if and only if $G \cong K_n$.

Theorem 5.2.3 ([9]) *Let G be a nontrivial connected graph with maximum degree $\Delta(G)$ and minimum degree $\delta(G)$. Then we have*

$$2\delta(G)H(G) \leq H_A(G) \leq 2\Delta(G)H(G)$$

with either equality holding if and only if G is a regular graph.

Theorem 5.2.4 ([9]) *Let G be a connected graph of order n. Then we have*

$$H_A(G) \leq DD(G)$$

with equality holding if and only if $G \cong K_n$.

Theorem 5.2.5 ([9]) *Suppose that G is a connected graph of order n and with m edges and diameter d. Then we have*

$$\frac{2(n-1)m}{d} + \frac{d-1}{d}M_1(G) \leq H_A(G) \leq (n-1)m + \frac{M_1(G)}{2}$$

with either equality holding if and only if $d \leq 2$.

Now, we turn to the formulae of additively weighted Harary indices of some graph products. The signs of some graph products follow those in Sect. 4.1.

Theorem 5.2.6 ([8]) *Let G_i be a connected graph of order n_i and with m_i edges for $i = 1, 2$. Then we have*

$$H_A\left(G_1 \bigvee G_2\right) = \frac{1}{2}\left(M_1(G_1) + M_1(G_2)\right) + (n_1 + n_2 - 1)(m_1 + m_2)$$

$$+ \frac{1}{2}n_2n_2(3n_1 + 3n_2 - 2) + 2(n_2m_1 + n_1m_2).$$

In [12, 21], the authors also independently gave a formula for $H_A(G_1 \bigvee G_2)$, but in it the first Zagreb co-indices $\overline{M_1}(G_i)$ for $i = 1, 2$ are also involved. Moreover, the corollaries below can be easily obtained.

Corollary 5.2.7 ([12]) $H_A(K_{s,t}) = s^2t + st^2 + s\binom{t}{2} + t\binom{s}{2}.$

Corollary 5.2.8 ([8]) *Let G_i be a connected graph of order n_i and with m_i edges for $i = 1, 2, \ldots, k$. Set $n = \sum_{i=1}^{k} n_i$ and $G = G_1 \bigvee G_2 \bigvee \cdots \bigvee G_k$. Then, we have*

$$H_A(G) = \sum_{i=1}^{k}\left(\frac{1}{2}M_1(G_i) + (n-1)m_i + (n-n_i)\binom{n_i}{2}\right)$$

$$+ \sum_{1 \leq i < j \leq k}\left(n_in_j(2n - n_i - n_j) + 2n_ie_j + 2n_je_i\right).$$

Denote by $K_{n_1,n_2,\ldots,n_k}$ a complete k-partition graph with a partition of cardinalities $n_1, n_2, \ldots, n_k$, respectively. The formula of $H_A(K_{n_1,n_2,\ldots,n_k})$ is determined in the following corollary.

Corollary 5.2.9 ([8])

$$H_A(K_{n_1,n_2,\ldots,n_k}) = \sum_{i=1}^{k}(n-n_i)\binom{n_i}{2} + \sum_{1 \leq i < j \leq k} n_in_j(2n - n_i - n_j).$$

Fig. 5.3 The graphs W_7 and F_7

Corollary 5.2.10 ([8]) *Let G be a connected graph of order n and with m edges. Then, we have*

$$H_A\left(G \bigvee K_1\right) = \frac{1}{2}\left(M_1(G) + 2(n+2)m + 3n^2 + n\right).$$

Recall that the wheel graph is $W_{n+1} \cong C_n \bigvee K_1$ and the fan graph is just $F_{n+1} \cong P_n \bigvee K_1$. For example, the graphs W_7 and F_7 are shown in Fig. 5.3. Moreover, the following remark can be easily deduced from Corollary 5.2.10.

Remark 5.2.11 $H_A(W_{n+1}) = \frac{5}{2}n^2 + \frac{9}{2}n, \quad H_A(F_{n+1}) = \frac{5}{2}n^2 + \frac{7}{2}n - 5.$

Theorem 5.2.12 ([8]) *Let G_i be a connected graph of order n_i and with m_i edges for $i = 1, 2$. Then we have*

$$H_A(G_1[G_2]) = n_2^2 H_A(G_1) + 4n_2 m_2 H(G_1) + \frac{1}{2}n_1 M_1(G_2)$$
$$+ 2m_1 n_2\left(\binom{n_2}{2} + m_2\right) + n_1 m_2(n_2 - 1).$$

The fence graph and closed fence [8] are $P_n[K_2]$ and $C_n[K_2]$, respectively. For example, $P_5[K_2]$ and $C_5[K_2]$ are shown in Fig. 5.4.

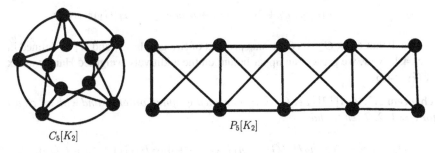

$C_5[K_2]$

$P_5[K_2]$

Fig. 5.4 The graphs $P_5[K_2]$ and $C_5[K_2]$

Corollary 5.2.13 ([8])

$$H_A(P_n[K_2]) = \sum_{i=1}^{k}(n-n_i)\binom{n_i}{2} + \sum_{1\leq i<j\leq k} n_i n_j(2n-n_i-n_j)$$

$$H_A(C_n[K_2]) = \sum_{i=1}^{k}(n-n_i)\binom{n_i}{2} + \sum_{1\leq i<j\leq k} n_i n_j(2n-n_i-n_j).$$

Theorem 5.2.14 ([8]) *Let G_i be a connected graph of order n_i and with m_i edges for $i = 1, 2$. Set $\overline{m_i} = \binom{n_i}{2} - m_i$ for $i = 1, 2$. Then*

$$H_A(G_1 \otimes G_2) = (n_2^3 - 4n_2m_2)M_1(G_1) + (n_1^3 - 4n_1m_1)M_1(G_2) + M_1(G_1)M_1(G_2)$$
$$+ \frac{1}{2}\Big[(n_2^2 + 2n_2\overline{m_2})\overline{M_1}(G_1) + (n_1^2 + 2n_1\overline{m_1})\overline{M_1}(G_2)\Big]$$
$$+ 8n_1n_2m_1m_2 + 2(n_1m_2\overline{m_1} + n_2m_1\overline{m_2}) - \frac{1}{2}\overline{M_1}(G_1)\overline{M_1}(G_2).$$

For two given graphs G_1 and G_2, the *symmetric difference* $G_1 \oplus G_2$ [8] is the graph with vertex set $V(G_1) \times V(G_2)$ and edge set

$$E(G_1 \oplus G_2) = \{(u_1, u_2)(v_1, v_2) | u_1v_1 \in E(G_1) \text{ or } u_2v_2 \in E(G_2) \text{ but not both.}$$

The equation for the Harary index of symmetric difference $G_1 \oplus G_2$ is presented in the following theorem.

Theorem 5.2.15 ([8]) *Let G_i be a connected graph of order n_i and with m_i edges for $i = 1, 2$. Set $\overline{m_i} = \binom{n_i}{2} - m_i$ for $i = 1, 2$. Then*

$$H_A(G_1 \oplus G_2) = (n_2^3 - 7n_2m_2)M_1(G_1) + (n_1^3 - 7n_1m_1)M_1(G_2) + 2M_1(G_1)M_1(G_2)$$
$$+ \frac{1}{2}\Big[(n_2^2 + 2n_2\overline{m_2} - 4m_2)\overline{M_1}(G_1) + (n_1^2 + 2n_1\overline{m_1} - 4m_1)\overline{M_1}(G_2)\Big]$$
$$+ 8n_1n_2m_1m_2 + 2(n_1m_2\overline{m_1} + n_2m_1\overline{m_2}) - \overline{M_1}(G_1)\overline{M_1}(G_2).$$

Recall that $G_1 \times G_2$ is a Cartesian product of two graphs G_1 and G_1 as defined in Sect. 4.1. Now, we present an upper bound on the additively weighted Harary index of $G_1 \times G_2$.

Theorem 5.2.16 ([12]) *Let G_i be a connected graph of order n_i and with m_i edges for $i = 1, 2$. Then we have*

$$H_A(G_1 \times G_2) \leq n_2 H_A(G_1) + n_1 H_A(G_2) + 4m_2 H(G_1) + 4m_1 H(G_2).$$

For two graphs G_1 and G_2, the *strong product* of G_1 and G_2, denoted by $G_1 \boxtimes G_2$, is a graph with vertex set $V(G_1) \times V(G_2)$ and two vertices (u, v) and (x, y) are adjacent if and only if

(i) $u = v$ and $xy \in E(G_2)$, or
(ii) $uv \in E(G_1)$ and $x = y$, or
(iii) $uv \in E(G_1)$ and $xy \in E(G_2)$.

In the theorem below, we give a formula for the additively weighted Harary index of $G \boxtimes K_r$.

Theorem 5.2.17 ([21]) *Let G be a connected graph of order n and with m edges. Then we have*

$$H_A(G \boxtimes K_r) = r^3 H_A(G) + 2r^2(r - 1)H(G) + 2r^2(r - 1)m + nr(r - 1)^2.$$

Since two graphs $G[K_r]$ and $G \boxtimes K_r$ are isomorphic, we can get the following corollary immediately.

Corollary 5.2.18 *For any connected graph G of order n and with m edges, we have*

$$H_A(G[K_r]) = r^3 H_A(G) + 2r^2(r - 1)H(G) + 2r^2(r - 1)m + nr(r - 1)^2.$$

We end this section with the following theorem, in which the formula of additively weighted Harary index is given for the corona product of two graphs.

Theorem 5.2.19 ([21]) *Let G_i be a graph of order n_i and with m_i edges for $i = 1, 2$. Then we have*

$$H_A(G_1 \circ G_2) = n_2^3 H_A(G_1) + 2H(G_1)\left(2m_2 + M_1(G_2) + \overline{M_1}(G_2)\right)$$
$$+ n_2 m_1(n_2^2 + 2m_2 - n_2) + \frac{n_1}{2}\left(2M_1(G_2) + \overline{M_1}(G_2)\right).$$

References

1. Lučić B, Miličević A, Nikolić S, Trinajstić N (2002) Harary index-twelve years later. Croat Chem Acta 75:847–868
2. Lučić B, Sović I, Plavšić D, Trinajstić N (2012) Harary matrices: definitions, properties and applications. In: Gutman I, Furtula B (eds) Distance in molecular graphs-applications. University of Kragujevac, Kragujevac, pp 3–26
3. Todeschini R, Consonni V (2000) Handbook of molecular descriptors. Wiley-VCH, Weinheim
4. Todeschini R, Consonni V (2009) Molecular descriptors for chemoinformatics, vol I, vol II. Wiley-VCH, Weinheim, pp 934–938
5. Das KC, Xu K, Cangul IN, Cevik AS, Graovac A (2013) On the Harary index of graph operations. J Inequal Appl 2013:1–16
6. Ricotta C, Stanisci A, Avena GC, Blasi C (2000) Quantifying the network connectivity of landscape mosaics: a graph-theoretical approach. Community Ecol 1:89–94

7. Jordán F, Báldi A, Orci KM, Rácz I, Varga Z (2003) Characterizing the importance of habitat patches and corridors in the maintenance of landscape connectivity of a Pholidoptera transsylvanica (Orthoptera) metapopulation. Landsc Ecol 18:83–92

8. Alizadeh Y, Iranmanesh A, Došlić T (2013) Additively weighted Harary index of some composite graphs. Discret Math 313:26–34

9. Hua H, Zhang S (2012) On the reciprocal degree distance of graphs. Discret Appl Math 160:1152–1163

10. Dobrynin AA, Kochetova AA (1994) Degree distance of a graph: a degree analogue of the Wiener index. J Chem Inf Comput Sci 34:1082–1086

11. Gutman I (1994) Selected properties of the Schultz molecular topogical index. J Chem Inf Comput Sci 34:1087–1089

12. Su G, Xiong L, Su X, Chen X, Some results on the reciprocal sum-degree distance of graphs. J Comb Optim. doi:10.1007/s10878-013-9645-5

13. Xu K, Klavžar S, Das KC, Wang J (2014) Extremal (n, m)-graphs with respect to distance-degree-based topological indices. MATCH Commun Math Comput Chem 72:865–880

14. Li S, Meng X, Four edge-grafting theorems on the reciprocal degree distance of graphs and their applications. J Comb Optim. doi:10.1007/s10878-013-9649-1

15. Sedlar J (2013) Extremal unicyclic graphs with respect to additively weighted Harary index. arXiv:1306.416v1(math.co)

16. Deng H, Krishnakumari B, Venkatakrishnan YB, Balachandran S, Multiplicatively weighted Harary index of graphs. J Comb Optim. doi:10.1007/s10878-013-9698-5

17. Li S, Zhang H, Some extremal properties of the multiplicatively weighted Harary index of a graph. J Comb Optim. doi:10.1007/s10878-014-9802-5

18. Xu K, Wang J, Das KC, Klavžar S (2014) Weighted Harary indices of apex trees and k-apex trees. Manuscript

19. Ashrafi AR, Došlić T, Hamzeh A (2010) The Zagreb coindices of graph operations. Discret Appl Math 158:1571–1578

20. Došlić T (2008) Vertex-weighted Wiener polynomials for composite graphs. Ars Math Contemp 1:66–80

21. Pattabiraman K, Vijayaragavan M (2013) Reciprocal degree distance of some graph operations. Trans Comb 2:13–24

Chapter 6
Open Problems

Despite many interesting results on the Harary index and additively or multiplicatively weighted Harary index of graphs reported in the previous chapters, there are still some complicated and challenging problems, which remain open to us, on this topic. In this chapter, we will propose various open problems on the Harary index and additively or multiplicatively weighted Harary index of graphs, most of them seem attractive to us and worthy of further research in the future.

6.1 Determining the Minimal Harary Index in a Given Set

In general, it is more difficult to characterize the extremal graphs with minimal Harary index in a given set of graphs than ones with maximal graphs. In this section, we will list some related problems to the minimal Harary index of graphs.

In Theorems 2.1.14 and 2.1.16, we have characterized completely the extremal graphs with maximal Harary indices among all nontrivial quasi-trees of order n and all k-generalized quasi-trees of order n, respectively. Moreover in Theorem 2.1.15, we determined the extremal graph with minimal Harary index among all 2-generalized quasi-trees of order n. However, it seems more intractable to determine the minimal Harary index of all k-generalized quasi-trees of order n for $k \geq 3$. This occurs mainly because that, when $k \geq 3$, the k-generalized quasi-trees have a more complicated structure than nontrivial quasi-trees. Therefore, we will point out the following problem as a research task in the future.

Problem 6.1.1 Characterizing the graphs from $QT^{(k)}(n)$ with minimal Harary index for $k \geq 3$ where $QT^{(k)}(n)$ is the set of k-generalized quasi-tree graphs of order n.

In [1], we have characterized the extremal graph with maximal Harary index among all connected graphs of order n and with k edges. But it seems a bit difficult to find the extremal one with minimal Harary index.

© The Author(s) 2015
K. Xu et al., *The Harary Index of a Graph*,
SpringerBriefs in Mathematical Methods, DOI 10.1007/978-3-662-45843-3_6

Problem 6.1.2 Finding the extremal graph with minimal Harary index among all connected graphs of order n and with k edges.

Recall the respective definitions of Wiener index, hyper-Wiener index, and the sign $\gamma(G, k)$ in Sect. 1.3, we can obtain the following equalities:

$$W(G) = \sum_{k \geq 1} k\gamma(G, k), \tag{6.1}$$

$$WW(G) = \sum_{k \geq 1} \left(\frac{k}{2} + \frac{k^2}{2} \right) \gamma(G, k). \tag{6.2}$$

Combining the formulae (1.2), (6.1), and (6.2) of Harary index, Wiener index, and hyper-Wiener index, respectively, they have a very analogous form. Considering this fact and the results in [2] on extremal self-complementary graphs with respect to Wiener index, we would like to present the following problem.

Problem 6.1.3 Find the extremal self-complementary graphs of order n with respect to Harary index, especially the minimal case.

More results with respect to distance-based topological indices can be found in [3]. Considering them, we can deal with the extremal results with respect to Harary index, especially the minimal case. Nevertheless, we can also consider the extremal graphs with respect to additively weighted Harary index (H_A) and multiplicatively weighted Harary index (H_M) from various sets of graphs. Maybe we will obtain some unexpected results compared to those on Harary index.

6.2 Other Attractive Open Problems

In this section, we report some other open problems on the Harary index of graphs.

As stated in Sect. 6.1, Harary index, Wiener index, and hyper-Wiener index have a high analogousness in form. In [4], we have proposed the following problems:

Problem 6.2.1 In what set $\mathcal{G}(n)$ of connected graphs of order n, the extremal (maximal or minimal) Harary index and the extremal (minimal or maximal) hyper-Wiener index are attained at the same graph?

Problem 6.2.2 In what set $\mathcal{G}(n)$ of connected graphs of order n, the extremal (maximal or minimal) Harary index and the extremal (minimal or maximal) Wiener index are attained at the same graph?

Problem 6.2.3 Which are the further relations among these three topological indices: Wiener index, hyper-Wiener index, and Harary index, especially between (hyper-) Wiener index and Harary index?

For Problems 6.2.1 and 6.2.2, we can find some positive results in [3]. But it seems a bit difficult to characterize the sets of graphs in which all graphs satisfy the property given in Problems 6.2.1 and 6.2.2.

Considering the respective definitions of additively and multiplicatively weighted Hararies indices of graphs, a natural problem below occurs in our mind:

Problem 6.2.4 In what set $\mathcal{G}(n)$ of connected graphs of order n, the extremal (maximal or minimal) additively and the extremal (maximal or minimal) multiplicatively weighted Harary index are attained at the same graph?

From Theorems 5.1.9, 5.1.16, 5.1.14, 5.1.21, 5.1.25, 5.1.26, 5.1.27, and 5.1.28, we get positive results to this problem. But the answer for other general sets of graphs are still unknown to us.

As pointed out in [5], a *dumbbell* $D(n, a, b)$ consists of a path P_{n-a-b} together with a independent vertices adjacent to one pendant vertex of the path and b independent vertices adjacent to the other pendant vertex of this path. For example, $D(13, 3, 5)$ is shown in Fig. 6.1. The dumbbell $D(n, a, b)$ is said to be a *balanced dumbbell* if $|a - b| \leq 1$.

It was shown [6] that the balanced dumbbell $D(n, \lceil \frac{2\alpha - n + 1}{2} \rceil, \lfloor \frac{2\alpha - n + 1}{2} \rfloor)$ has the maximal Wiener index among all trees of order n and with independence number α. Inspired by the above result and Problem 6.2.2, initially we thought that the same dumbbell has the minimal Harary index among all trees with n vertices and independence number α. But this is actually false since we have given a counterexample to it in [5]. Therefore, we would like to propose, again, the following challenging problem.

Problem 6.2.5 Characterizing the extremal tree with minimal Harary index among trees of order n and independence number α.

For Problem 6.2.3 some nice results have been obtained in [7–9]. However, it is worthy of further research on this topic, especially on the relation between Harary index and (hyper-) Wiener index of graphs.

In [5], we have presented a relationship between Harary index and first and second Zagreb indices for trees (see Theorem 3.2.5). However, we are totally unknown to this relation between them for general graphs. Here, we would like to provide the readers of this book with the following interesting problem.

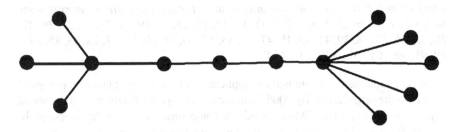

Fig. 6.1 The dumbbell $D(13, 3, 5)$

Problem 6.2.6 Finding some nice relationship between Harary index and Zagreb index for general graphs, or for some special graphs.

In Sect. 3.1, we have obtained two necessary conditions (Theorems 3.1.3 and 3.1.4) and a sufficient one (Corollary 3.1.5) for a graph G satisfying $H(G) = RCW(G)$. But these conditions are merely on some special graphs. Here, we would like to propose the following problem on this topic:

Problem 6.2.7 Finding a necessary and sufficient condition for general graphs G with $RCW(G) = H(G)$.

But, to our best knowledge, this problem seems a bit difficult. At least at this present stage, to solve it is impossible. Maybe characterizing the mathematical correlation of these two topological indices will be helpful to this problem.

Additively weighted Harary index and multiplicatively weighted Harary index [10] are two fundamental variants of Harary index of graphs, some results on the former one have been reported in Chap. 5. But there are very few results on the latter one.

Similar to Problem 6.2.3, we give a problem on additively weighted Harary index and multiplicatively weighted Harary index as follows:

Problem 6.2.8 Which are the further relations between additively weighted Harary index and multiplicatively weighted Harary index of some graphs, even of some special ones?

In analogy with Problem 6.1.1, considering the results in [11], naturally we propose the following problem on the additively and multiplicatively weighted Harary indices of graphs.

Problem 6.2.9 Characterizing the graphs from $\mathcal{QT}^{(k)}(n)$ with minimal additively and multiplicatively weighted Harary indices, respectively, for $k \geq 3$ where $\mathcal{QT}^{(k)}(n)$ is the set of k-generalized quasi-tree graphs of order n.

In 1998, Lepovic and Gutman [12] first proposed the inverse problem for Wiener index of trees as follows:

Given a positive integer a, find whether there exists a tree such that $W(T) = a$? Moreover, they also presented the following related conjecture:

Conjecture 6.2.10 *Every positive integer not in the set A is a Wiener index of some tree, where $A = \{2, 3, 5, 6, 7, 8, 11, 12, 13, 14, 15, 17, 19, 21, 22, 23, 24, 26, 27, 30, 33, 34, 37, 38, 39, 41, 43, 45, 47, 51, 53, 55, 60, 61, 69, 73, 77, 78, 83, 85, 87, 89, 91, 99, 101, 106, 113, 147, 159\}.*

In [13], Goldman et al. studied the application of inverse problem of topological indices to the drug design. By 2006, Wang and Yu [14] completely proved the above conjecture. In [15], Li and Wang solved two conjectures in [13] on the inverse problems of Wiener indices of peptoids, one of which is proved and the other is disproved. Furthermore, Li et al. [16] studied the inverse problems of other topological indices, including the first Zagreb index (M_1), Merrifield–Simmons index, Hosoya index (see in [17] a recent extremal result with respect to them), and so on.

Considering the close relation between Wiener index and Harary index, we would give an interesting inverse problem for the Harary index as follows:

Problem 6.2.11 Given a positive rational number p, find whether there exists a connected graph G such that $H(G) = p$?

It seems more difficult to solve Problem 6.2.11. Maybe we can consider a special case when p is in a given interval $[a, b]$. Anyway, Problem itself is a very attractive topic to us. More exactly, we will list the following problem which can be viewed, more or less as an inverse problem for the Harary index.

Problem 6.2.12 Given a set $\mathcal{G}(n)$ of graphs with $m \leq H(G) \leq M$, for $a = pm + qM$ with $p + q = 1$ and $0 < p, q < 1$, find a graph G such that $H(G) = a$.

In addition to those two above, similar inverse problem with respect to Additively weighted Harary index or multiplicatively weighted Harary index can be another direction for us in the near future.

Considering the chemical background of Harary index [18, 19] when it was first introduced in 1993, we would like to end this book with the following interesting problem.

Problem 6.2.13 Find some new (potential) applications of Harary index in Organic Chemistry or Mathematical Chemistry.

References

1. Xu K, Trinajstić N (2011) Hyper-Wiener and Harary indices of graphs with cut edges. Util Math 84:153–163
2. Hendry GRT (1989) On mean distance in certain classes of graphs. Networks 19:451–457
3. Xu K, Liu M, Das KC, Gutman I, Furtula B (2014) A survey on graphs extremal with respect to distance-based topological indices. MATCH Commun Math Comput Chem 71:461–508
4. Xu K (2012) Trees with the seven smallest and eight greatest Harary indices. Discret Appl Math 160:321–331
5. Das KC, Xu K, Gutman I (2013) On Zagreb and Harary indices. MATCH Commun Math Comput Chem 70:301–314
6. Dankelmann P (1994) Average distance and independence number. Discret Appl Math 51:75–83
7. Gutman I (2002) Relation between hyper-Wiener and Wiener index. Chem Phys Lett 364:352–356
8. Gutman I, Furtula B (2003) Hyper-Wiener index vs. Wiener index. Two highly correlated structure-descriptors. Monatshefte Math 134:975–981
9. Zhou B, Gutman I (2004) Relations between Wiener, hyper-Wiener and Zagreb indices. Chem Phys Lett 394:93–95
10. Alizadeh Y, Iranmanesh A, Došlić T (2013) Additively weighted Harary index of some composite graphs. Discret Math 313:26–34
11. Xu K, Wang J, Das KC, Klavžar S (2014) Weighted Harary indices of apex trees and k-apex trees, manuscript

12. Lepovic M, Gutman I (1998) A collective property of trees and chemical trees. J Chem Inf Comput Sci 38:823–826
13. Goldman D, Istrail S, Piccolboni GLA (2000) Algorithmic strategies in combinatorial chemistry. In: Proceedings of the 11th ACM-SIAM symposium on discrete algorithms, pp 275–284
14. Wang H, Yu G (2006) All but 49 numbers are Wiener indices of trees. Acta Appl Math 92:15–20
15. Li X, Wang L (2003) Solutions for two conjectures on the inverse problem of the Wiener index of peptoids. SIAM J Discret Math 17:210–218
16. Li X, Li Z, Wang L (2003) The inverse problems for some topological indices in combinatorial chemistry. J Comput Biol 10:47–55
17. Xu K, Li J, Zhong L (2012) The Hosoya indices and Merrifield-Simmoms indices of graphs with connectivity at most k. Appl Math Lett 25:476–480
18. Ivanciuc O, Balaban TS, Balaban AT (1993) Design of topological indices. Part 4. Reciprocal distance matrix, related local vertex invariants and topological indices. J Math Chem 12:309–318
19. Plavšić D, Nikolić S, Trinajstić N, Mihalić Z (1993) On the Harary index for the characterization of chemical graphs. J Math Chem 12:235–250

Printed in the United States
By Bookmasters